The Topological Dynamics of Ellis Actions

MEMOIRS
of the
American Mathematical Society

Number 913

The Topological Dynamics
of Ellis Actions

Ethan Akin
Joseph Auslander
Eli Glasner

September 2008 • Volume 195 • Number 913 (end of volume) • ISSN 0065-9266

American Mathematical Society
Providence, Rhode Island

2000 *Mathematics Subject Classification.* Primary 20M20, 37B05, 37B20, 54H20.

Library of Congress Cataloging-in-Publication Data
Akin, Ethan, 1946–
 The topological dynamics of Ellis actions / Ethan Akin, Joseph Auslander, Eli Glasner.
  p. cm. — (Memoirs of the American Mathematical Society, ISSN 0065-9266 ; no. 913)
 "Volume 195, number 913 (end of volume)."
 Includes bibliographical references and index.
 ISBN 978-0-8218-4188-4 (alk. paper)
  1. Topological transformation groups. 2. Topological semigroups. I. Auslander, Joseph, 1930– II. Glasner, Eli, 1945– III. Title.

QA613.7.A438   2008
514—dc22   2008021012

Memoirs of the American Mathematical Society

This journal is devoted entirely to research in pure and applied mathematics.

Subscription information. The 2008 subscription begins with volume 191 and consists of six mailings, each containing one or more numbers. Subscription prices for 2008 are US$675 list, US$540 institutional member. A late charge of 10% of the subscription price will be imposed on orders received from nonmembers after January 1 of the subscription year. Subscribers outside the United States and India must pay a postage surcharge of US$38; subscribers in India must pay a postage surcharge of US$43. Expedited delivery to destinations in North America US$53; elsewhere US$130. Each number may be ordered separately; *please specify number* when ordering an individual number. For prices and titles of recently released numbers, see the New Publications sections of the *Notices of the American Mathematical Society*.

Back number information. For back issues see the *AMS Catalog of Publications*.

Subscriptions and orders should be addressed to the American Mathematical Society, P. O. Box 845904, Boston, MA 02284-5904, USA. *All orders must be accompanied by payment.* Other correspondence should be addressed to 201 Charles Street, Providence, RI 02904-2294, USA.

Copying and reprinting. Individual readers of this publication, and nonprofit libraries acting for them, are permitted to make fair use of the material, such as to copy a chapter for use in teaching or research. Permission is granted to quote brief passages from this publication in reviews, provided the customary acknowledgment of the source is given.

Republication, systematic copying, or multiple reproduction of any material in this publication is permitted only under license from the American Mathematical Society. Requests for such permission should be addressed to the Acquisitions Department, American Mathematical Society, 201 Charles Street, Providence, Rhode Island 02904-2294, USA. Requests can also be made by e-mail to reprint-permission@ams.org.

Memoirs of the American Mathematical Society (ISSN 0065-9266) is published bimonthly (each volume consisting usually of more than one number) by the American Mathematical Society at 201 Charles Street, Providence, RI 02904-2294, USA. Periodicals postage paid at Providence, RI. Postmaster: Send address changes to Memoirs, American Mathematical Society, 201 Charles Street, Providence, RI 02904-2294, USA.

© 2008 by the American Mathematical Society. All rights reserved.
This publication is indexed in *Science Citation Index*®, *SciSearch*®, *Research Alert*®, *CompuMath Citation Index*®, *Current Contents*®/*Physical, Chemical & Earth Sciences*.
Printed in the United States of America.

∞ The paper used in this book is acid-free and falls within the guidelines
established to ensure permanence and durability.
Visit the AMS home page at http://www.ams.org/

10 9 8 7 6 5 4 3 2 1   13 12 11 10 09 08

Contents

Introduction		1
Chapter 1.	Semigroups, Monoids and Their Actions	9
Chapter 2.	Ellis Semigroups and Ellis Actions	19
Chapter 3.	Continuity Conditions	35
Chapter 4.	Applications Using Ideals	45
Chapter 5.	Classical Dynamical Systems	59
Chapter 6.	Classical Actions: The Group Case	87
Chapter 7.	Classical Actions: The Abelian Case	121
Chapter 8.	Iterations of Continuous Maps	135
Table		143
Bibliography		145
Index		149

Abstract

An Ellis semigroup is a compact space with a semigroup multiplication which is continuous in only one variable. An Ellis action is an action of an Ellis semigroup on a compact space such that for each point in the space the evaluation map from the semigroup to the space is continuous. At first the weak linkage between the topology and the algebra discourages expectations that such structures will have much utility. However, Ellis has demonstrated that these actions arise naturally from classical topological actions of locally compact groups on compact spaces and provide a useful tool for the study of such actions. In fact, via the apparatus of the enveloping semigroup the classical theory of topological dynamics is subsumed by the theory of Ellis actions. Our exposition describes and extends Ellis' theory and demonstrates its usefulness by unifying many recently introduced concepts related to proximality and distality. Moreover, our approach leads us to several results which are new even in the classical setup.

Received by the editor September 29, 2003, Revised October 31, 2005.

2000 *Mathematics Subject Classification.* Primary 20M20, 37B05, 37B20, 54H20.

Key words and phrases. Ellis semigroup, enveloping semigroup, dynamical systems, recurrence, minimal, distal, proximal, asymptotic, weak-mixing.

Introduction

A classical discrete time dynamical system is a pair (X, f) consisting of a nonempty, compact, Hausdorff state space X together with a continuous function $f : X \to X$, which we will call a map *on* X. For example, if we begin with a smooth vectorfield on a compact manifold and integrate to get the flow then the time 1 map defines a discrete time dynamical system.

We will call the system *metric* when the state space X is metrizable as well as compact. The system is *reversible* when f is a homeomorphism in which case (X, f^{-1}) is the *reverse system*.

We are primarily interested in the long term behavior of the system. For an initial point $x \in X$ we denote by $\omega f(x)$ the limit point set of the *orbit sequence* $\{x, f(x), f^2(x), ...\}$ so that

$$(0.1) \qquad \omega f(x) \quad =_{def} \quad \bigcap_n \overline{\{f^i(x) : i \geq n\}}.$$

Thus, the *orbit closure* of x consists of the union of the orbit of x with $\omega f(x)$.

Because X is compact, $\omega f(x)$ is never empty. This is one of the typical applications of the assumptions of compactness. While the assumption of metrizability is sometimes convenient, e.g. for applications of the Baire Category Theorem, the requirement of compactness is often essential. Even when the initial focus of study is on a noncompact state space Y, e.g. $\mathbb{R}^n$, we tend to reduce to the compact case, either by restricting to some compact invariant subset or by compactifying the system, i.e. by embedding Y in a compact space X to which the dynamics extends. The Stone-Čech compactification βY of a completely regular space Y is a functorial construction and so is a natural choice for compactification. The only disadvantage is that βY is usually quite large. When Y is metrizable we often prefer to use some metrizable compactification for the system, which exists if Y has a countable base.

When the orbit of x is dense in X we call x a *transitive point* and denote by $TRANS_f$ the set of transitive points. If f is a surjective map then x is a transitive point iff $\omega f(x) = X$. We call the system topologically point transitive or just *point transitive* when $TRANS_f \neq \emptyset$. We call the system *minimal* if $TRANS_f = X$, i.e. every point is transitive. The system is minimal iff $\omega f(x) = X$ for all $x \in X$.

A subset $A \subset X$ is *+ invariant* (or *invariant*) if $f(A) \subset A$ (resp. $f(A) = A$). To a nonempty, closed, + invariant subset is associated the *subsystem* $(A, f|A)$, which we will usually write as (A, f). For example, $\omega f(x)$ is a nonempty, closed, invariant subset.

A subset A is called a *point transitive subset* (a *minimal subset*) if it is a nonempty, closed, + invariant subset whose associated subsystem is transitive (resp. minimal). A point x is called a *minimal point* if it lies in some minimal subset of X, or, equivalently $\omega f(x)$ is a minimal subset and $x \in \omega f(x)$.

In general, we call x a *recurrent point* when $x \in \omega f(x)$. This is equivalent to saying that x lies in a nonempty, closed, invariant subset with x a transitive point for the associated subsystem.

Of special interest is a sharpening of the condition transitivity. (X, f) is called *weak mixing* when the product system $(X \times X, f \times f)$ is transitive. It then follows by a theorem of Furstenberg that the system is *weak mixing of all orders* meaning that the products $(X^n, f \times .... \times f)$ are transitive for all positive integers n.

We supplement the study of individual orbits by comparing the behaviors of different orbits. A pair of points $(x, y) \in X \times X$ is called *asymptotic* if their orbits are eventually indistinguishable. In the metric case this says that for every positive ϵ the points $f^n(x)$ and $f^n(y)$ are eventually ϵ close. If for every positive ϵ the orbits are frequently ϵ close then we call the pair *proximal*. We denote by $ASYMP$ and by $PROX$ the set of asymptotic pairs and the set of proximal pairs, respectively. The subset $ASYMP \subset X \times X$ is clearly an equivalence relation, but $PROX$ need not be. Nonetheless, we denote by $PROX(x)$ the set $\{y : (x, y) \in PROX\}$ and call it the *proximal cell* of x. A point x is called *distal* if it is proximal to no other point in its orbit closure, or, equivalently, if $\omega f(x) \cap PROX(x) = \{x\}$. If distal points occur, then the system is called *point distal*. It is called *distal* when every point is distal.

If (X, f) and (Y, g) are discrete time dynamical systems and $\pi : X \to Y$ is a continuous map, then π is called a dynamical systems homomorphism or an *action map* if it maps f to g, i.e. $\pi \circ f = g \circ \pi$. It then follows that the closed equivalence relation

$$(0.2) \qquad R_\pi \;=_{def}\; \{(x_1, x_2) : \pi(x_1) = \pi(x_2)\} \;=\; (\pi \times \pi)^{-1}(1_Y)$$

is a + invariant subset of $X \times X$ and so defines a subsystem of $(X \times X, f \times f)$. Here 1_Y is the diagonal subset of $Y \times Y$, i.e. the identity function. The subsystem of the product $(R_\pi, f \times f)$ is used to extend ideas from systems to action maps between systems. Notice that exactly when Y is a singleton, R_π is all of $X \times X$.

As an example of how these concepts are used, let us glance at the heavily trodden area of chaos theory.

For a dynamical system (X, f), consisting of a surjective continuous map on a compact metric space, there are a number of definitions for the term "chaos". Associated with Devaney is sensitive dependence upon initial conditions. Many authors prefer the stronger assumption of positive entropy.

Another definition is implicit in Li and Yorke's original paper (1975). They call a subset $D \subset X$ *scrambled* if for every pair of orbits with initial points $x \neq y \in D$ the sequence of distances $\{d(f^n(x), f^n(y))\}$ neither approaches zero nor is bounded away from zero. That is, the pair (x, y) is proximal but not asymptotic for the system (X, f). They call a system chaotic when it contains an uncountable scrambled set. We will use the term *Li-Yorke chaotic*.

Clearly, if a non-diagonal pair (x, y) is recurrent for $(X \times X, f \times f)$ then it is not asymptotic. We will call $D \subset X$ *strongly scrambled* if every nondiagonal pair in $D \times D$ is proximal and recurrent. We will call the system *strongly Li-Yorke chaotic* when it admits an uncountable strongly scrambled subset.

We associate these notions with subsets of $X \times X$. For $(x, y) \in X \times X$ we write

$$(x, y) \in 1_X \iff x = y.$$
$$(x, y) \in PROX \iff \omega(f \times f)(x, y) \cap 1_X \neq \emptyset.$$
$$(x, y) \in ASYMP \iff \omega(f \times f)(x, y) \subset 1_X.$$
$$(x, y) \in RECUR^2 \iff (x, y) \in \omega(f \times f)(x, y).$$
$$(x, y) \in LI - YORKE \iff (x, y) \in 1_X \cup (PROX \setminus ASYMP).$$
$$(x, y) \in sLI - YORKE \iff (x, y) \in PROX \cap RECUR^2.$$

The set $sLI - YORKE$ like $PROX$ and $RECUR^2$ is a G_δ while $LI - YORKE$ is usually not so. This makes the stronger notion technically easier to study in a number of ways.

For minimal systems the relations among the different notions is now understood. In general, strong Li-Yorke chaos implies Li-Yorke chaos and Li-Yorke chaos does not occur in equicontinuous systems. Auslander and Yorke (1986) proved that a minimal system is either sensitive or equicontinuous. So for minimal systems Li-Yorke chaos implies sensitive dependence. On the other hand, Blanchard et al. (2002) prove that for minimal systems positive entropy implies Li-Yorke chaos and the same proof yields strong Li-Yorke chaos. Huang and Ye (2000) prove that a minimal system admits a dense scrambled set iff the system is weak mixing. Blanchard et al. (2002) show that there is then a scrambled set which is a dense, countable union of Cantor Sets in X. Again it can be chosen strongly scrambled.

Complementary to all this is distality. The system (X, f) is *distal* if $PROX = 1_X$. Blanchard et al. (2002) call the system *almost distal* if $LI-YORKE = 1_X$. We call the system *semidistal* if $sLI-YORKE \subset 1_X$.

In each case, there is an action map version, defined using R_π. The action map $\pi : (X, f) \to (Y, g)$ is called distal/almost distal/semidistal if $R_\pi \cap PROX/R_\pi \cap LI - YORKE/R_\pi \cap sLI - YORKE$ is contained in the diagonal 1_X where $R_\pi = (\pi \times \pi)^{-1}(1_Y)$. The system or space concept reduces to the map concept with (Y, g) trivial, i.e. Y is a singleton.

By contrast, π is called a *proximal* map (or an *asymptotic* map) if $R_\pi \subset PROX(f)$ (resp. $R_\pi \subset ASYMP(f)$).

The above discussion can be rephrased in terms of actions. By letting 1 act via f a discrete time classical dynamical system (X, f) can be regarded as an action of the monoid $\mathbb{Z}_+$. If the system is reversible then this extends to an action of the group $\mathbb{Z}$. A flow is similarly an action of the monoid $\mathbb{R}_+$ or the group $\mathbb{R}$. The concept of dynamical system has now been routinely extended to include the action of any topological group G although at least local compactness is usually assumed.

We have seen that for any initial point x we often compactify the orbit of x by taking its closure, which adjoins $\omega f(x)$ to the positive orbit. This suggests the idea of compactifying the acting semigroups in order to close up all the orbits at once. Thus, we will study these matters by using the category of Ellis actions of an Ellis semigroup on general compact Hausdorff spaces. A semigroup is a set S equipped with an associative multiplication map $M : S \times S \to S$. An Ellis semigroup is a semigroup with a compact Hausdorff topology with respect to which the maps $p \mapsto pq = M(p,q)$ are continuous for each $q \in S$.

Some motivating examples are:
- The Stone-Čech compactification $\beta\mathbb{Z}_+$.
- More generally, a locally compact group G admits an associated right-invariant uniformity generated by the neighborhoods of the identity element. The Gelfand space of the Banach algebra of bounded, real-valued functions, uniformly continuous with respect to this uniformity naturally forms an Ellis semigroup which we denote $\beta_u G$.
- For a compact space X the space of maps X^X with the product topology.
- Given a compact dynamical system (X, f) the closure in X^X of the set $\{f^n : n \in \mathbb{Z}_+\}$, denoted $E = E(X, f)$, is the enveloping semigroup associated with (X, f), defined and extensively exploited by Robert Ellis.

This one-sided continuity appears, at first glance, to be a rather weak foundation upon which to build the structure of topological dynamics. In particular, the Ellis semigroups are usually not topological semigroups for which the multiplication M itself is continuous.

Contrast the situation with some Lie group results. In Tondeur's notes (1965) it is observed that for a Lie group G the tangent bundle TG is a Lie group. The result is left to the reader with the hint that the functor T conserves products. The point is that the associative law and other group axioms can be described by various commutative diagrams using the multiplication $M : G \times G \to G$. These just lift to show that $TM : TG \times TG \to TG$ is a group multiplication. Furthermore, the base space projection and the zero-section map are natural transformations between the identity functor and T. It follows that $\tau : TG \to G$ and $j_0 : G \to TG$ are group homomorphisms. The splitting by j_0 of the short exact sequence $T_eG \to TG \to G$ expresses TG as a semi-direct product via the adjoint action of G on T_eG, which is the Lie algebra. We thus obtain a parallelization of the manifold G. The associated constant sections are the invariant vectorfields. That original exercise with hint takes you pretty far.

One of us, reading this in graduate school, was inspired to use the Stone-Čech compactification functor in a similar fashion. Alas, β does not conserve products. From the continuous map $\beta M : \beta(G \times G) \to \beta G$ it is not possible to get a continuous group action on βG. By hacking around you can construct an associative multiplication on βG, but the result is not continuous. The whole enterprise seemed barren. Others have probably tried the same thing and given up at the same point. Meanwhile, Robert Ellis had for more than ten years been demonstrating that the resulting structure is immensely fruitful. In fact, the book which collected together his results, Ellis (1969), was published only a couple of years after that grad school encounter with Tondeur. This algebraic approach was used by Ellis in his discovery of a Galois theory of distal extensions and a new proof of Furstenberg's structure theorem. Later this theory was applied by several authors to develop a structure theory for general minimal dynamical systems (or flows), see Veech (1970), (1977), Ellis-Glasner-Shapiro (1975) and MacMahon (1976). Even for $\beta \mathbb{Z}_+$ alone there is an elaborate and beautiful structure. The current work is summarized in the impressive treatise Hindman and Strauss (1998). In addition, a parallel theory has developed, which was motivated by functional analysis instead of dynamics, see Burckel (1970) and Ruppert (1984).

Initially no topology is needed. In Chapter 1 we consider the category of semigroup actions on sets. For such an action $\Phi : S \times X \to X$ we can define recurrence, proximality and the various notions of distality. For example, $x \in X$ is a *recurrent point* when $x \in Sx$. Since we do not assume that S has an identity, not every point need be recurrent. The point x is minimal when it is contained in a minimal set (= a minimal invariant subset), or, equivalently, when x is recurrent and its orbit Sx is minimal. A point x is a *transitive point* when $Sx = X$. The system Φ is called *point transitive* if it admits a transitive point.

Even at this level there are important results. For example, the composition of distal (or semidistal) action maps is distal (resp. semidistal) without any topological assumptions. The analogous result for almost distal does not hold even for classical discrete time metric systems.

These results highlight by contrast the deeper results where topology is required.

In Chapter 2 we introduce Ellis actions of Ellis semigroups. Φ is an *Ellis action* if S is an Ellis semigroup, X is a compact Hausdorff space and $p \mapsto px$ is continuous for for each $x \in X$. From the compactness and continuity assumptions we obtain the existence of various sorts of minimal invariant, closed subsets. Of special importance is the Ellis-Namakura demonstration of the existence of idempotent elements in any Ellis semigroup. For Ellis actions we obtain semigroup characterizations of the three sorts of distality. There are also parallel X^2 results. Call X *point minimal* when every point is minimal, ie. X is a union of minimal subsets. It is *orbit minimal* when every orbit is a minimal set and it is *recurrent point minimal* when every recurrent point is minimal. The system Φ is distal/almost distal/semidistal iff the product action on $X \times X$ is point minimal/orbit minimal/recurrent point minimal.

The multiplication $M : S \times S \to S$ for an Ellis semigroup is itself an Ellis action which we call the *translation action* on S. The closed invariant subsets with respect to this action are the closed (left) ideals of S and the minimal elements of S are thus the elements of the minimal ideals.

In Chapter 3 we consider equicontinuity for Ellis actions. For an Ellis action $\Phi : S \times X \to X$ the maps $\Phi^p : X \to X$ defined by $x \mapsto px$ are usually not continuous. We call Φ equicontinuous when the set $\{\Phi^p : p \in S\}$ is equicontinuous. Φ is equicontinuous iff it is a topological action, i.e. iff Φ is continuous. In turn this is equivalent to equicontinuity for a subset $\{\Phi^p : p \in D\}$ with D a dense subset of S. This is, of course, a very strong condition. We call Φ *densely continuous* if for p in a dense subset of S each map Φ^p is continuous. This much weaker condition is applied in the later sections.

The unifying power of the above trinity of distality concepts is extended in Chapter 4. We obtain additional concepts by applying the earlier definitions to the Ellis actions of closed subsemigroups of the Ellis semigroup S. For example, let $[Min(S)]$ denote the smallest closed invariant subset of S which contains the minimal elements. Distality for S and for $[Min(S)]$ agree, but the system is $[Min(S)]$ almost distal iff the proximality relation on X is an equivalence relation and a minimal system is $[Min(S)]$ semidistal iff it satisfies a condition which in the classical case characterizes those minimal systems which are so-called PI systems.

All this illustrates the flexibility of this abstract, Ellis actions approach. Even when our main interest is in applications to classical actions, we obtain useful results by applying the general theory to the Ellis actions obtained by restricting to such closed subsemigroups.

Furthermore, suppose that $\Phi : S \times X \to X$ and $\Psi : S \times Y \to Y$ are Ellis actions. A continuous map $\pi : X \to Y$ is an action map if $\pi(px) = p\pi(x)$ for all $(p, x) \in S \times X$. If Ψ is assumed to be a minimal action then many results about the action map π can be reduced to space results for the action of Iso_y on $\pi^{-1}(y)$ where Iso_y is the closed subsemigroup $\{p \in S : py = y\}$.

In Chapter 5 we return to the classical systems. A triple (S, G, S^*) is called a *classical Ellis semigroup* when G is a dense subsemigroup such that the multiplication map $M : G \times S \to S$ is continuous and such that the closed subset $S^* \subset S$ is a closed two-sided ideal with $S = G \cup S^*$. Usually, but not always, $S^* = S \setminus G$. The motivating example has $S = \beta\mathbb{Z}_+$, $G = \mathbb{Z}_+$ and $S^* = \beta^*\mathbb{Z}_+ =_{def} \beta\mathbb{Z}_+ \setminus \mathbb{Z}_+$.

Φ is a classical Ellis action when the restriction $\Phi : G \times X \to X$ is continuous. In particular, Φ is densely continuous. It is assumed that S is monoid, i.e. it has a two-sided identity e which is contained in G but which is usually not in S^*. So for recurrence definitions we use the action of S^*.

In this context we introduce transitivity and other residual concepts, extending Akin and Glasner (2001). We show that a minimal semidistal classical system is disjoint from all scattering systems, extending the almost distality result from Blanchard et al. (2002). There are special results for classical actions when the semigroup G is either a group or is abelian. These are considered in Chapters 6 and 7, respectively.

When G is a not necessarily abelian group, the concept of weak mixing raises some problems. The Furstenberg Intersection Lemma implies that in the abelian case weak mixing implies weak mixing of all orders and this is not true in the nonabelian case. We introduce the concept of adjoint transitivity and obtain a nonabelian version of the Furstenberg Intersection Lemma. This implies that a classical system with G a group is weak mixing of all orders iff it is weak mixing and adjoint transitive.

Here again our abstract approach yields new results even for the classical case. As an example let us mention the fact, obtained in Chapter 6, that an asymptotic classical action map $\pi : X \to Y$ between metric minimal systems is necessarily almost one-to-one.

Finally, in Chapter 8 we return to the situation described in our introductory remarks, where a map f induces a classical Ellis action of $(\beta\mathbb{Z}_+, \mathbb{Z}_+, \beta^*\mathbb{Z}_+)$ on X. The associated classical Ellis action definitions specialize to the original ones. We construct examples of some almost distal systems for which various closely related systems are not almost distal. For example, if (X, f) is semidistal or distal with f a homeomorphism then (X, f^{-1}) satisfies the corresponding property. We construct an example where (X, f) is an asymptotic lift of a distal system and so is almost distal but the reverse (X, f^{-1}) is not almost distal.

Another example is that of an iteration of two asymptotic extensions of minimal systems $X \xrightarrow{\pi} Z \xrightarrow{\rho} Y$ such that the composition $\rho \circ \pi : X \to Y$ is almost one-to-one but not asymptotic. It then follows that when Y is distal X is semidistal but not almost distal.

We conclude by proving that a minimal semidistal system has topological entropy zero.

CHAPTER 1

Semigroups, Monoids and Their Actions

In this section we consider the definitions and results which require no topological assumptions and so which live in the category of sets and maps. For sets S, X we let X^S denote the set of all maps from S to X.

We will follow the notation of Akin (1997). In particular, notice that our actions occur on the left.

We write for a map $\Phi : S \times X \to X$

(1.1)
$$\begin{aligned}
px &= \Phi(p,x) = \Phi^p(x) = \Phi_x(p) \quad \text{for } (p,x) \in S \times X. \\
AB &= \{px : p \in A \text{ and } x \in B\} \quad \text{for } A \times B \subset S \times X. \\
\Phi^\# &: S \to X^X \quad \text{is defined by } p \mapsto \Phi^p. \\
\Phi_\# &: X \to X^S \quad \text{is defined by } x \mapsto \Phi_x.
\end{aligned}$$

A semigroup S is a nonempty set equipped with $M : S \times S \to S$ which is an associative multiplication, i.e.

(1.2) $\quad\quad M^p \circ M^q = M^{pq} \quad \text{for all } p, q \in S.$

An action of a semigroup S on a nonempty set X is a map $\Phi : S \times X \to X$ which is an action, i.e.

(1.3) $\quad\quad \Phi^p \circ \Phi^q = \Phi^{pq} \quad \text{for all } p, q \in S.$

We will often speak, somewhat abusively, of *the system* X meaning the set X equipped with an understood semigroup action Φ.

If S is a semigroup then the multiplication map M is an action of S on itself, called the *translation action*. If X is a singleton set then the unique map from $S \times X$ to X is an action called the *trivial action*.

A semigroup is called a *monoid* when it contains a two-sided identity element, u, i.e. $up = p = pu$ for all $p \in S$. Such an identity is unique. A *monoid action* of a monoid S on a set X is a semi-group action such that the identity element acts as the identity map on X, i.e. $\Phi^u = 1_X$. The translation action of a monoid is a monoid action as is the trivial action of any monoid.

For any semigroup S we define the *monoid extension* S^+ by

(1.4) $\quad\quad S^+ =_{def} S \cup \{e\}$

where e is a point not in S and the multiplication is extended by $ep = p = pe$ for all $p \in S^+$. A semigroup action of S on X extends uniquely to a monoid action of S^+ on X.

9

If S, T are semigroups then $g : S \to T$ is a *semigroup homomorphism* when $g(pq) = g(p)g(q)$ for all $p, q \in S$. A nonempty subset $H \subset S$ is a *subsemigroup* when it is closed under multiplication, i.e. $HH \subset H$, in which case the inclusion of H into S is a homomorphism. The image $g(H) \subset T$ is a subsemigroup and the preimage of a subsemigroup of T is a subsemigroup of S when it is nonempty. A singleton $\{u\}$ is a subsemigroup iff u is an *idempotent*, i.e. $uu = u$. A semigroup action of T on X can be pulled back via the homomorphism g to define a semigroup action of S on X. To be precise: if $\Phi : T \times X \to X$ is a T action then $\Phi \circ (g \times 1_X) : S \times X \to X$ is an S action which is the *pullback of Φ via g*. In particular, S acts, via g, on T. When the homomorphism is the inclusion of a subsemigroup H we speak of the *restriction* of the action to H. We write $\Phi|H$ for the restriction of the action to H, i.e. the restriction of the map Φ to $H \times X$.

If S, T are monoids then $g : S \to T$ is a *monoid homomorphism* when it is a semigroup homomorphism which maps the identity of S to that of T. If S is a monoid with identity u and $g : S \to T$ is a surjective semigroup homomorphism then $g(u)$ is an identity in T. That is, T is a monoid and g is a monoid homomorphism. A subsemigroup of a monoid S is a *submonoid* when it contains the identity element of S. A semigroup homomorphism extends uniquely to a monoid homomorphism of the monoid extensions.

If $\Phi : S \times X \to X$ and $\Psi : S \times Y \to Y$ are semigroup actions, then $\pi : X \to Y$ is an *action map* when $\pi(px) = p\pi(x)$ for all $p \in S, x \in X$. A subset $K \subset X$ is called *invariant* if $K \neq \emptyset$ and $SK \subset K$. In that case, the restricted map $\Phi|K : S \times K \to K$ is a semigroup action and the inclusion of K into X is an action map. We call $\Phi|K$ or just K a *subsystem* of X. The image $\pi(K) \subset Y$ is then invariant and the preimage of an invariant subset of Y is an invariant subset of X if it is nonempty. The *orbit* of a point $x \in X$ is the invariant set $Sx = \Phi_x(S)$. Any union of invariant sets is invariant and any nonempty intersection of invariant sets is invariant.

If $\Phi : S \times X \to X$ is any semigroup action then there is a unique action map of Φ onto any trivial action. We call such a map a *trivial action map*.

An *ideal* in a semigroup S is a subset $K \subset S$ which is invariant with respect to the translation action, i.e. $SK \subset K$. For $p \in S$ the orbit under translation, Sp, is an ideal in S called the *principal ideal associated with p*. If $g : S \to T$ is a homomorphism and $L \subset T$ is an ideal then the preimage $g^{-1}(L)$ is an ideal if it is nonempty. If g is surjective then the image of an ideal in S is an ideal in T. Any union of ideals is an ideal and any nonempty intersection of ideals is an ideal.

Thus, an ideal $K \subset S$ is a subsemigroup of a special sort. On the other hand, a subsemigroup H of S is called a *co-ideal*, or sometimes in the literature a *cancelation subsemigroup*, if $pq \in H$ and $q \in H$ imply $p \in H$.

For any nonempty set X map composition gives X^X the structure of a monoid and the evaluation map $Ev : X^X \times X \to X$ is a monoid action. If $\Phi : S \times X \to X$ is a semigroup action then $\Phi^\# : S \to X^X$ is a homomorphism and for each $x \in X$, $\Phi_x = Ev_x \circ \Phi^\# : S \to X$ is an action map whose image is the orbit of x. In particular, if S is a semigroup and $p \in S$ then the translation map $M_p : S \to S$ defined by $q \mapsto qp$ is an action map with image the principal ideal Sp.

If a semigroup S acts on each member of an indexed family $\{X_i : i \in I\}$ then the *product action* on the product $\Pi\{X_i : i \in I\}$ is uniquely defined by the condition that for every $j \in I$ the projection map $\pi_j : \Pi\{X_i : i \in I\} \to X_j$ is an action map. In particular, if $\Phi : S \times X \to X$ is an action and I is any set then we obtain the product action $\Phi^I : S \times X^I \to X^I$. We denote by Φ^2 the product action on $X \times X$, i.e. the special case with $I = \{0, 1\}$. When $I = X$ we have $(\Phi^X)_{1_X} = \Phi^\#$.

The most important example of a surjective homomorphism is the restriction $\Phi^\# : S \to \Phi^\#(S) \subset X^X$. We call $\Phi^\#(S)$ the *enveloping semigroup* for the action Φ. As a subsemigroup of X^X it acts on X with

$$(1.5) \qquad px \quad = \quad \Phi^\#(p)(x).$$

We can often use this to replace S by the enveloping semigroup. The action of $\Phi^\#(S)$ is *faithful*, i.e. $px = qx$ for all $x \in X$ implies $p = q$ in $\Phi^\#(S)$.

For $\Phi : S \times X \to X$ a semigroup action and for $x, x_1, x_2 \in X$ we define the *isotropy set* for the point x and the *focus set* for the pair (x_1, x_2):

$$(1.6) \qquad \begin{aligned} Iso_x &=_{def} (\Phi_x)^{-1}(x) = \{p \in S : px = x\} \\ Foc_{(x_1,x_2)} &=_{def} (\Phi^2_{(x_1,x_2)})^{-1}(1_X) = \{p \in S : px_1 = px_2\}. \end{aligned}$$

$Foc_{(x_1,x_2)}$ is an ideal if it is nonempty and Iso_x is a co-ideal if it is nonempty. By (1.5) the preimage of $Iso_x \subset \Phi^\#(S)$ is $Iso_x \subset S$ with a similar result for $Foc_{(x_1,x_2)}$.

We call x a *recurrent point* if $Iso_x \neq \emptyset$ and a *fixed point* if $Iso_x = S$. Thus, x is a recurrent point iff it lies in its own orbit, i.e. $x \in Sx$ and it is a fixed point iff it is the only point in its orbit, i.e. $\{x\} = Sx$. At the opposite extreme, we call x a *transitive point* if the its orbit is the entire space, i.e. $X = Sx$ and we denote by $TRANS$ the set of transitive points so that

$$(1.7) \qquad TRANS \quad =_{def} \quad \{x \in X : Sx = X\}.$$

The system is called *point transitive* when $TRANS$ is nonempty. Clearly, any transitive point is recurrent.

An invariant subset $M \subset X$ is called *minimal* when it is minimal with respect to the ordering by inclusion in the set of invariant subsets of X. Recall that an invariant subset is assumed to be nonempty. Since the orbit Sx is an invariant set contained in any invariant set which contains x, it follows that M is a minimal subset iff $M = Sx$ for all $x \in M$. In particular, the points of a minimal subset are recurrent. A point $x \in X$ is called a *minimal point* when it is an element of minimal subset. Clearly, x is a minimal point iff it is recurrent and Sx is a minimal subset. That is, x is minimal iff $x \in Sx$ and $y \in Sx$ implies $Sy = Sx$. Finally, we say that Φ or X is *minimal* when X itself is a minimal subset. Thus, X is minimal iff

$Sx = X$ for all $x \in X$, i.e. iff $TRANS = X$. So X is minimal when it has no proper invariant subset.

Since the intersection of two invariant subsets is invariant if it is nonempty, it follows that any two distinct minimal subsets are disjoint. The union of the minimal subsets of X is denoted $Min(X)$. Thus, $x \in Min(X)$ iff x is a minimal point of X.

If $\Phi : S \times X \to X$ and $\Psi : S \times Y \to Y$ are semigroup actions and $\pi : X \to Y$ is an action map then $M \subset X$ minimal implies $\pi(M) \subset Y$ is minimal.

An ideal $J \subset S$ is a minimal ideal when it is a minimal invariant set with respect to the translation action on S, in which case $J = Sp = Jp$ for any $p \in J$. $Min(S)$ is the union of all the minimal ideals of S. If it is nonempty then $Min(S)$ an ideal consisting of all minimal elements of S.

We use the term *ideal* to refer to a left ideal. A *two-sided ideal* has certain additional properties.

Lemma 1.1. Let S be a semigroup.

(a) If $Min(S)$ is nonempty then it is a two-sided ideal. That is, $S \cdot Min(S) \cdot S \subset Min(S)$.
(b) Any two-sided ideal J of S contains all the minimal points. That is, $S \cdot J \cdot S \subset J$ implies $Min(S) \subset J$.
(c) S itself is minimal iff for every $p \in S$, $M_p : S \to S$ is a surjective map.
(d) S is a group iff for every $p \in S$, $M_p, M^p : S \to S$ are both surjective maps.

PROOF. (a): We saw above that $Min(S)$ as the union of the minimal ideals is an ideal if it is nonempty. For any $p \in S$, $M_p : S \to S$ is an action map with respect to the translation action and so it maps minimal points to minimal points. Thus, $Min(S)$ is a right ideal as well.

(b): Let p be any minimal point. Since J is an ideal, Jp is an ideal contained in the minimal ideal Sp and so, by minimality, $Jp = Sp$. Since a minimal point is recurrent, $p \in Sp = Jp$. Finally, since J is a two-sided ideal, $Jp \subset J$.

(c): $Sp = S$ for all $p \in S$ iff every M_p is surjective.

(d): This is a classical algebra exercise. Start with $p_0 \in S$. There exists $\bar{p}_0 \in S$ such that $\bar{p}_0 p_0 p_0 = p_0$ because $M_{p_0 p_0}$ is surjective. Hence, for all $q \in S$, $\bar{p}_0 p_0 p_0 q = p_0 q$. Since M^{p_0} is surjective this implies $e_L p = p$ for all $p \in S$ with $e_L = \bar{p}_0 p_0$. Similarly, there exists a right identity e_R. $e_R = e_L e_R = e_L$ and so this common element e is both a left and right identity. For every, p there exist p_L and p_R such that $p_L p = e = p p_R$ and so $p_R = p_L p p_R = p_L$. That is, p has an inverse. That translations of a group are bijective is obvious. □

We call a pair (x_1, x_2) *proximal* if $Foc_{(x_1,x_2)} \neq \emptyset$ and *asymptotic* if $Foc_{(x_1,x_2)} = S$.

Using these subsets of the semigroup, we define

$$\begin{aligned}
RECUR^n \quad &=_{def} \quad \{z \in X^n : z \in Sz\} \\
&= \{z \in X^n : Iso_z \neq \emptyset\} \\
&= \{z \in X^n : pz = z \text{ for some } p \in S\}.
\end{aligned}$$
(1.8)

$$\begin{aligned}
PROX \quad &=_{def} \quad \{(x_1, x_2) \in X \times X : S(x_1, x_2) \cap 1_X \neq \emptyset\} \\
&= \{(x_1, x_2) \in X \times X : Foc_{(x_1,x_2)} \neq \emptyset\} \\
&= \{(x_1, x_2) \in X \times X : px_1 = px_2 \text{ for some } p \in S\}
\end{aligned}$$
(1.9)

$$\begin{aligned}
ASYMP \quad &=_{def} \quad \{(x_1, x_2) \in X \times X : S(x_1, x_2) \subset 1_X\} \\
&= \{(x_1, x_2) \in X \times X : Foc_{(x_1,x_2)} = S\} \\
&= \{(x_1, x_2) \in X \times X : px_1 = px_2 \text{ for all } p \in S\}
\end{aligned}$$
(1.10)

$$PROX \wedge RECUR \quad =_{def} \quad PROX \cap RECUR^2 \tag{1.11}$$

We will write $RECUR$ for $RECUR^1$. For $PPP = RECUR^n, PROX$, etc. we will write $PPP(\Phi)$ or $PPP(X)$ when we have to distinguish the action or space. When we have to distinguish the semigroup we will use subscripts, writing PPP_S. For example, if H is a subsemigroup of S we write PPP_H for the subsets defined using the restriction of the action to H. From equation (1.5) it is easy to see that, in this notation,

$$PPP_{\Phi^\#(H)} \quad = \quad PPP_H \quad = \quad PPP(\Phi|H) \tag{1.12}$$

for $PPP = RECUR^n, PROX$, etc. Thus, when studying these sets, we can replace the action of S by the faithful action of the enveloping semigroup $\Phi^\#(S)$ when it is convenient to do so.

Clearly, only a diagonal pair can be both asymptotic and recurrent. That is,

$$ASYMP \cap RECUR^2 \quad \subset \quad 1_X. \tag{1.13}$$

For a monoid action all points are recurrent and only diagonal points are asymptotic. In particular, for the monoid extension of the S action:

$$\begin{aligned}
RECUR^n_{S+} \quad &= \quad X^n. \\
(PROX \wedge RECUR)_{S+} \quad &= \quad PROX_{S+} \quad = \quad PROX_S. \\
ASYMP_{S+} \quad &= \quad 1_X.
\end{aligned}$$
(1.14)

We use these sets to define various notions of distality and proximality for systems.

Definition 1.2. Let $\Phi : S \times X \to X$ be a semigroup action. We call the action Φ or the system X

> *trivial* when $X \times X \quad \subset \quad 1_X$.
> *proximal* when $X \times X \quad \subset \quad PROX$.
> *asymptotic* when $X \times X \quad \subset \quad ASYMP$.

distal when $PROX \subset 1_X$.
almost distal when $PROX \subset ASYMP$.
semidistal when $PROX \wedge RECUR \subset 1_X$.

Notice that in the first five cases we can replace the inclusions by equalities as the reverse inclusions always hold.

Each of these definitions has a map version which extends the space concept. Let $\Phi : S \times X \to X$ and $\Psi : S \times Y \to Y$ be semigroup actions and $\pi : X \to Y$ be an action map. Let $X_y =_{def} \pi^{-1}(y)$ for $y \in Y$, which is nonempty iff y is in the image $\pi(X) \subset Y$. Define an invariant subset of $R_\pi \subset X \times X$, and a map $\lambda_\pi : R_\pi \to Y$.

$$\begin{aligned}R_\pi &=_{def} (\pi \times \pi)^{-1}(1_Y) = \bigcup\{X_y \times X_y : y \in Y\} \\ &= \{(x_1, x_2) : \pi(x_1) = \pi(x_2)\} \subset X \times X. \\ \lambda_\pi(x_1, x_2) &=_{def} \pi(x_1) = \pi(x_2).\end{aligned}$$
(1.15)

λ_π is an action map from $\Phi^2|R_\pi$ to Ψ. By identifying the invariant subset $1_Y \subset Y \times Y$ with Y itself we can identify λ_π with $\pi \times \pi$.

When y is a recurrent point of $\pi(X) \subset Y$, we denote by $\Phi(y) : Iso_y \times X_y \to X_y$ the Iso_y action on X_y which is the restriction of the map Φ.

When π is the trivial action map then $R_\pi = X \times X$.

Definition 1.3. Let $\Phi : S \times X \to X$ and $\Psi : S \times Y \to Y$ be semigroup actions and $\pi : X \to Y$ be an action map. We call the action map π

injective when $R_\pi \subset 1_X$
proximal when $R_\pi \subset PROX$.
asymptotic when $R_\pi \subset ASYMP$.
distal when $R_\pi \cap PROX \subset 1_X$.
almost distal when $R_\pi \cap PROX \subset ASYMP$.
semidistal when $R_\pi \cap PROX \cap RECUR^2 \subset 1_X$.

Clearly, each of these properties holds for the action $\Phi : S \times X \to X$ iff it holds for the action map π from X to the trivial action. On the other hand, the following result will sometimes allow us to obtain map consequences from space results.

Proposition 1.4. Let $\Phi : S \times X \to X$ and $\Psi : S \times Y \to Y$ be semigroup actions and $\pi : X \to Y$ be an action map. Assume that $y \in \pi(X) \subset Y$ is a recurrent point. If the action map π is asymptotic, distal, almost distal or semidistal then the Iso_y action on the fiber X_y, $\Phi(y) : Iso_y \times X_y \to X_y$, satisfies the corresponding property.

PROOF. If a pair $z = (x_1, x_2) \in X_y \times X_y$ is in $ASYMP \subset X \times X$ then it is obviously asymptotic with respect to $Iso_y \subset S$. Thus, π asymptotic implies $\Phi(y)$ is asymptotic. If the pair z is proximal with respect to Iso_y then $z \in R_\pi \cap PROX$. If, in addition, z is recurrent with respect to Iso_y then $z \in R_\pi \cap PROX \cap RECUR^2$. In the first case, π distal will imply $x_1 = x_2$ while in the second case π semidistal will imply $x_1 = x_2$.

Finally, if the pair z is Iso_y proximal and π is almost distal then z is asymptotic in $X \times X$ and so is Iso_y asymptotic. Thus, $\Phi(y)$ is almost distal. □

These properties are related as follows.

(1.16)
$$\begin{array}{rcl} \text{injective} & \Longrightarrow & \text{asymptotic and distal.} \\ \text{asymptotic} & \Longrightarrow & \text{proximal and almost distal.} \\ \text{distal} & \Longrightarrow & \text{almost distal} \Longrightarrow \text{semidistal.} \end{array}$$

N.B. For monoid actions the three distality concepts agree. An action map π of monoid actions is distal iff it is almost distal iff it is semidistal. In particular, for any action map of S actions these three concepts collapse together for the map of the S^+ extended actions. On the other hand, (1.14) implies that the concepts of distality, proximality and, of course, injectivity are the same for maps of S actions and for their S^+ extensions. We will repeatedly use this to obtain results about distality from the corresponding semidistality results applied to the S^+ extension.

We will use the following result most often when the homomorphism g is the inclusion of a subsemigroup.

Lemma 1.5. Let $\Phi : S \times X \to X$ and $\Psi : S \times Y \to Y$ be semigroup actions, $\pi : X \to Y$ be an action map and $g : T \to S$ be a semigroup homomorphism.
- (a) If π is asymptotic, distal, almost distal or semidistal as an action map of S actions then the corresponding property holds for π as an action map of the pulled back T actions.
- (b) If π is a proximal action map for the pulled back T actions then it is a proximal map of S actions.
- (c) Assume that every ideal J in S meets the image $g(T)$, or, equivalently, the preimage $g^{-1}(J)$ of every ideal J in S is nonempty. π is a distal or proximal action map of S actions iff the corresponding property holds for π as an action map of the pulled back T actions.

PROOF. Observe first that

(1.17)
$$\begin{array}{rcl} RECUR_T & \subset & RECUR_S. \\ PROX_T & \subset & PROX_S. \end{array}$$

While, on the other hand,

(1.18) $$ASYMP_S \subset ASYMP_T.$$

Finally, given the assumption in (c), $Foc_{(x_1,x_2)} \neq \emptyset$ iff $g^{-1}(Foc_{(x_1,x_2)}) \neq \emptyset$. Thus, if every ideal of S meets $g(T)$ then

(1.19) $$PROX_T \;=\; PROX_S.$$

From these relations the results are easy exercises. $\square$

Lemma 1.6. Let $\Phi : S \times X \to X$ and $\Psi : S \times Y \to Y$ be semigroup actions and $\pi : X \to Y$ be an action map.

(a) The following inclusions hold:

(1.20)
$$\begin{aligned} \pi(RECUR(X)) &\subset RECUR(Y) \\ (\pi \times \pi)(PROX(X)) &\subset PROX(Y). \\ (\pi \times \pi)(ASYMP(X)) &\subset ASYMP(Y). \\ (\pi \times \pi)(PROX \wedge RECUR(X)) &\subset PROX \wedge RECUR(Y). \end{aligned}$$

and

(1.21) $\quad \pi$ surjective $\quad \Longrightarrow \quad \pi(TRANS(X)) \subset TRANS(Y)$.

(b) If π is injective then

(1.22)
$$\begin{aligned} (\pi)^{-1}(RECUR(Y)) &= RECUR(X) \\ (\pi \times \pi)^{-1}(PROX(Y)) &= PROX(X). \\ (\pi \times \pi)^{-1}(ASYMP(Y)) &= ASYMP(X). \\ (\pi \times \pi)^{-1}(PROX \wedge RECUR(Y)) &= PROX \wedge RECUR(X). \end{aligned}$$

(c) The action map π is proximal iff $(\pi \times \pi)^{-1}(PROX(Y)) = PROX(X)$.

PROOF. (a),(b): If $x \in X$ and $y = \pi(x)$ then it is easy to check that

(1.23) $$\begin{aligned} \pi(Sx) &= Sy \quad \text{and} \\ Iso_x &\subset Iso_y, \end{aligned}$$

with equality when π is injective. Similarly, when $z \in X \times X$ and $w = (\pi \times \pi)(z)$

(1.24) $$Foc_z \;\subset\; Foc_w,$$

with equality when π is injective. These easily imply (1.20), (1.21) and (1.22).

(c): Assume π is proximal and that $z \in X \times X$ with $w = (\pi \times \pi)(z)$ a proximal pair. There exists $p \in S$ such that $pw \in 1_Y$ and so $pz \in R_\pi$. Since π is proximal there exists $q \in S$ such that $qpz \in 1_X$ and so z is a proximal pair.

Since $1_Y \subset PROX(Y)$, $R_\pi \subset (\pi \times \pi)^{-1}(PROX(Y))$. Thus, the reverse implication is clear. □

Proposition 1.7. Let $\Phi : S \times X \to X$, $\Psi : S \times Y \to Y$ and $\Theta : S \times Z \to Z$ be semigroup actions and let $X \xrightarrow{\pi} Y \xrightarrow{\epsilon} Z$ be action maps.
   (a) If the composition $\epsilon \circ \pi$ is an injective, asymptotic, proximal, distal, almost distal or semidistal map then π satisfies the corresponding property.
   (b) Assume that π is surjective. If the composition $\epsilon \circ \pi$ is an injective, asymptotic, or proximal map then ϵ satisfies the corresponding property.
   (c) Assume that π is injective. If ϵ is an injective, asymptotic, proximal, distal, almost distal or semidistal map then $\epsilon \circ \pi$ satisfies the corresponding property.
   (d) If both ϵ and π are either injective, proximal, distal or semidistal maps then the composition $\epsilon \circ \pi$ satisfies the corresponding property.

PROOF. (a),(b),(c): For (a) use
$$R_\pi \subset R_{\epsilon \circ \pi} = (\pi \times \pi)^{-1} R_\epsilon. \tag{1.25}$$
If π is surjective then
$$(\pi \times \pi) R_{\epsilon \circ \pi} = R_\epsilon. \tag{1.26}$$
This implies (b) and for (c) use, in addition, Lemma 1.6(b).

(d): If both maps are injective then the composition is. If both maps are proximal then by Lemma 1.6(c) and the equation in (1.25), the composition is proximal. Assume that both maps are semidistal, $z \in R_{\epsilon \circ \pi} \cap PROX \cap RECUR^2(X)$ and $w = (\pi \times \pi)(z)$. By (1.25) and Lemma 1.6(a), $w \in R_\epsilon \cap PROX \cap RECUR^2(Y)$ and so $w \in 1_Y$ since ϵ is semidistal. Hence, $z \in R_\pi$ and so $z \in 1_X$ since π is semidistal. This shows that the composition is semidistal. The result for distality follows from the semidistality result applied to the S^+ monoid extension. □

REMARK 1.1. The major result of this section is part (d) above. We will see that the composition of almost distal maps need not be almost distal.

Recall that if $\Phi : S \times X \to X$ is an action and $K \subset X$ is an invariant subset then we call the restricted S action $\Phi : S \times K \to K$ the *subsystem* K. Similarly, if $\pi : X \to Y$ is an action map, $K \subset X$ and $L \subset Y$ are invariant sets with $\pi(K) \subset L$ then the restriction $\pi| : K \to L$ of π is called the *subsystem* $\pi|$ of π.

Corollary 1.8. If an action map is asymptotic, proximal, distal, almost distal or semidistal then any subsystem satisfies the corresponding property.

PROOF. This follows from Proposition 1.7 but is easily proved directly. □

Proposition 1.9. If every member of an indexed family of action maps is an injective, asymptotic, distal, almost distal or semidistal map then the product map satisfies the corresponding property.

PROOF. This is an easy exercise applying Lemma 1.6(a) to the projection maps from the product to each factor. □

We conclude this chapter with some useful, albeit easy, results about the asymptotic relation

Proposition 1.10. Let $\Phi : S \times X \to X$ be a semigroup action.
  (a) The relation $ASYMP \subset X \times X$ is an invariant equivalence relation. For every $p \in S$ the map $\Phi^p : X \to X$ is constant on every $ASYMP$ equivalence class.
  (b) If $H \subset S$ is a subsemigroup which satisfies the condition $\overline{HS} \subset H$, i.e. H is a right ideal, then the relation $ASYMP_H$ is not merely H invariant but S invariant.

PROOF. Part (a) is obvious. In fact, for any subsemigroup H of S

$$(1.27) \qquad ASYMP_H = \bigcap_{p \in H} (\Phi^p \times \Phi^p)^{-1}(1_X).$$

(b): If $(x_1, x_2) \in ASYMP_H$ and $p \in S$ then for all $q \in H$, $qp \in H$ and so $qpx_1 = qpx_2$. Thus, $p(x_1, x_2) \in ASYMP_H$. That is, the set $ASYMP_H$ is invariant with respect to the product action $\Phi^2 : S \times X \times X \to X \times X$. □

CHAPTER 2

Ellis Semigroups and Ellis Actions

By a *space* we will mean a nonempty, compact, Hausdorff space. For spaces S, X the space X^S of all maps from S to X is equipped with the product topology. We consider Ellis semigroups and their actions as described in Ellis (1969) and Auslander (1988) and follow the notation of Akin (1997).

An *Ellis semigroup* is a space equipped with a semigroup multiplication M such that the adjoint map $M^\# : S \to S^S$ is continuous, or, equivalently, $M_q : S \to S$ is continuous for each $q \in S$. For any space X the space of maps X^X is an Ellis semigroup.

An *Ellis action* is a semigroup action Φ of an Ellis semigroup on a space X such that the semigroup homomorphism $\Phi^\# : S \to X^X$ is continuous, i.e. for each $x \in X$, the action map $\Phi_x : S \to X$ is continuous. The translation action and trivial actions of an Ellis semigroup are Ellis actions as is the evaluation action of X^X on the space X.

A closed subsemigroup H of an Ellis semigroup S is an Ellis semigroup in its own right. If $g : T \to S$ is a continuous homomorphism of Ellis semigroups then its image is a closed subsemigroup and the pullback via g of an Ellis action of S is an Ellis action of T. In particular, for an Ellis action Φ, the enveloping semigroup $\Phi^\#(S) \subset X^X$ is an Ellis semigroup and the restriction to it of the evaluation action pulls back via the continuous homomorphism $\Phi^\# : S \to X^X$ to the original Ellis action of S on X.

If $\Phi : S \times X \to X$ is an Ellis action then Iso_x is a closed subset of S for every $x \in X$. Hence, Iso_x is a closed co-ideal whenever it is nonempty, i.e. whenever the point x is recurrent.

Compactness is used at this level of the theory is to provide minimal elements in various classes of closed nonempty subsets.

Let $\Phi : S \times X \to X$ be an Ellis action. Recall that $M \subset X$ is called *minimal* when it is a minimal invariant subset. That is, M is an invariant subset and the restriction $\Phi|M$ is minimal, or, equivalently, $M = Sx$ for all $x \in M$. Since the orbit Sx is closed for an Ellis action, a minimal subset is a closed invariant subset and so $\Phi|M$ is an Ellis action. We can apply Zorn's Lemma to the set of closed invariant subsets. By compactness, for any collection of closed invariant sets which is linearly ordered by inclusion the intersection is nonempty and so is a closed invariant set. It follows that any closed invariant subset contains a minimal set. In fact, if K is any invariant subset and $x \in K$ the orbit Sx is a closed invariant subset contained in K. Hence, any invariant subset of X contains a minimal invariant subset.

In particular, for an Ellis action the set $Min(X)$ of all the minimal points of X is nonempty. In particular, for an Ellis semigroup S every ideal contains a minimal ideal which is necessarily closed in S and the union $Min(S)$ is nonempty ideal in S, usually not closed. If J is a minimal ideal and $p \in J$ then Jp and Sp are invariant subsets of J and so $Jp = Sp = J$ by minimality. In fact, if K is any other ideal then $Kp = J$.

If A is a nonempty subset of X then we denote by $[A]$ the smallest closed invariant subset of X which contains A, i.e. the intersection of all closed invariant subsets of X which contain A. Notice that since the action is not jointly continuous, the closure of an invariant set need not be invariant and while the closure of $S^+ A = A \cup SA$ is always contained in $[A]$, the inclusion might be strict.

If $\Phi : S \times X \to X$ and $\Psi : S \times Y \to Y$ are Ellis actions and $\pi : X \to Y$ is a continuous action map then $M \subset X$ minimal implies $\pi(M) \subset Y$ is minimal. This is just a semigroup result. On the other hand, if $K \subset Y$ is minimal and K meets, and so is contained in, the image $\pi(X) \subset Y$ then the invariant set $(\pi)^{-1}(K)$ contains minimal subsets. Any minimal subset M of $(\pi)^{-1}(K)$ maps onto K. In particular,

(2.1) $\qquad \pi$ surjective $\qquad \Longrightarrow \qquad \pi(Min(X)) \quad = \quad Min(Y)$.

An element $u \in S$ is *idempotent* if $u^2 = u$. For any $A \subset S$ we let $Id(A)$ denote the set of idempotents in A, i.e. $Id(A) =_{def} \{u \in A : u^2 = u\}$. We recall the following crucial property of an Ellis semigroup (see Ellis (1969) Lemma 2.9 and Auslander (1988) Lemma 6.6).

Lemma 2.1. Ellis-Numakura A closed subsemigroup A of an Ellis semigroup S contains idempotents.

PROOF. By Zorn's Lemma any closed subsemigroup contains a minimal member of the set of closed subsemigroups. So it suffices to show that a minimal closed subsemigroup K contains an idempotent. Let $p \in K$. Kp is a closed subsemigroup of K and so by minimality $Kp = K$. Hence, $p \in Kp$ which implies Iso_p meets K. Since $Iso_p \cap K$ is nonempty it is a closed subsemigroup of K and so by minimality again $Iso_p \cap K = K$. Hence, $p \in Iso_p$ which implies that p is an idempotent. In fact, since this implies that $\{p\}$ is a closed subsemigroup, $K = \{p\}$. □

If $g : S \to T$ is a surjective, continuous homomorphism of Ellis semigroups then the image of a minimal ideal is minimal and if $J \subset T$ is a minimal ideal then any minimal ideal in $g^{-1}(J)$ maps onto J. The image of an idempotent is idempotent and if $u \in T$ is an idempotent then any idempotent in the closed subsemigroup $g^{-1}(u)$ maps to u. Hence, if A is an

ideal of S, and B is a closed subsemigroup in S then
(2.2)
$$g(A \cap Min(S)) = g(A) \cap Min(T) \quad \text{and} \quad g(B \cap Id(S)) = g(B) \cap Id(T).$$

Lemma 2.2. Let J be an ideal and H be a closed co-ideal in an Ellis semigroup S.

(a) If $p \in S$ then Jp is an ideal in S which is minimal if J is.

(b) If $u \in Id(J)$ then the closed ideal Su is contained in J and for $p \in S$

(2.3)
$$pu = p \iff p \in Su.$$

(c) Let $u, v \in Id(S)$. The following conditions are equivalent and when they hold we write $u >_R v$

(2.4)
$$Sv \subset Su \iff v \in Su \iff vu = v.$$

The following conditions are equivalent and when they hold we write $u >_L v$

(2.5)
$$Iso_u \subset Iso_v \iff u \in Iso_v \iff uv = v.$$

Both $>_R$ and $>_L$ are quasi-orders on $Id(S)$, i.e. they are reflexive, transitive relations. Furthermore, an idempotent u is minimal iff it is minimal with respect to either quasi-order.

(d) Let $u, \tilde{v} \in Id(J)$. If $u >_R \tilde{v}$, or equivalently, $\tilde{v} \in Id(Su) \subset Id(J)$, then $v = u\tilde{v} \in Id(Su)$ and

(2.6)
$$vu = uv = v.$$

So that $u >_R \tilde{v} >_R v$ and $u >_L v$. If $\tilde{v}$ is a minimal idempotent then v is a minimal idempotent as well.

(e) Suppose that J is a minimal ideal such that $J \cap H \neq \emptyset$. Then:
  (i) $J \cap H$ is a minimal H ideal.
  (ii) Every ideal in S meets H.
  (iii) $Min(S) \cap H = Min(H)$.

PROOF. (a): Since M_p is an action map, $Jp = M_p(J)$ is minimal when J is.

(b): $p = qu$ implies $pu = quu = qu = p$.

(c),(d): We begin with the equivalences in (c).

Since Su is an ideal and $v \in Sv$ it is clear that $Sv \subset Su$ iff $v \in Su$. The final equivalence follows from (b).

Similarly, since Iso_v is a co-ideal and $u \in Iso_u$ it is clear that $Iso_u \subset Iso_v$ iff $u \in Iso_v$. The final equation is the definition of the condition $u \in Iso_v$.

It is clear that each relation is reflexive and transitive.

Now we prove (d).

As in (b), $v \in uSu$ implies (2.6). Hence,

(2.7)
$$vv = u\tilde{v}u\tilde{v} = u\tilde{v}\tilde{v} = u\tilde{v} = v.$$

So v is an idempotent. The relations $\tilde{v} >_R v$ and $u >_L v$ are clear.

If $\tilde{v}$ is minimal then v is minimal because it lies in the minimal ideal $S\tilde{v}$.

To complete the proof of (c), notice first that minimality of an idempotent is exactly $>_R$ minimality. With u such a minimal idempotent, let $v \in Id(S)$ with $u >_L v$, i.e $uv = v$. Let $w = vu$. We show that $w = u$ which says that $v >_L u$ and so implies that u is $>_L$ minimal. Note first that $uw = uvu = vu = w$. Next observe that the ideal Sw is contained in and so equals the minimal ideal Su. So $u \in Su = Sw$. By (b) applied to w, $uw = u$. Hence, $u = w$ as promised.

Finally, suppose that u is $>_L$ minimal. Choose $\tilde{v}$ a minimal idempotent in Su. By (d), already proved, $v = u\tilde{v}$ is a minimal idempotent with $u >_L v$. By assumption on u, $v >_L u$, i.e. $vu = u$. By part (a), the ideal $Su = (Sv)u$ is minimal because Sv is. Hence, $u \in Su$ is a minimal idempotent.

(e)(i): Because H is a subsemigroup and J is an ideal $H(J \cap H) \subset J \cap H$. Furthermore, if $q_1, q_2 \in J \cap H$ then minimality of J implies that $q_1 = pq_2$ for some $p \in S$. Because H is a co-ideal, $p \in H$. Thus, $Hq_2 = J \cap H$ for all $q_2 \in J \cap H$ and so $J \cap H$ is a minimal H ideal.

(e)(ii): Let K be an arbitrary ideal in S and let $q \in J \cap H$. Then Kq is an S ideal contained in J and so by minimality $Kq = J$. Hence, there exists $p \in K$ such that $pq = q$. Because H is a co-ideal, $p \in H$ and so K meets H.

(e)(iii): Let L be an H ideal. Notice first that

$$(2.8) \qquad SL \cap H \quad \subset \quad L$$

for if $h = sl$ with $h \in H, s \in S$ and $l \in L$ then because H is a co-ideal $s \in H$ and so $h \in HL \subset L$. Let K be a minimal S ideal contained in the S ideal SL. By (i) and (ii) $K \cap H$ is an H ideal and it is contained in L by (2.8). If, in addition, L is a minimal H ideal then $K \cap H = L$. Thus, every minimal H ideal is the intersection of H with some minimal S ideal, proving (iii). $\square$

Recall from Lemma 1.1 that

$$(2.9) \qquad S(Min(S))S \quad \subset \quad Min(S).$$

That is, while we have used the term *ideal* to denote a left ideal, $Min(S)$ is actually a two-sided ideal.

An idempotent u in an Ellis semigroup is called *maximal* if it is maximal with respect to the quasi-order $>_L$. That is, $u \in Id(S)$ is maximal iff

$$(2.10) \qquad v \in Id(S) \text{ and } vu = u \quad \Rightarrow \quad uv = v.$$

or, equivalently,

$$(2.11) \qquad v \in Id(S) \text{ and } Iso_v \subset Iso_u \quad \Rightarrow \quad Iso_v = Iso_u.$$

Proposition 2.3. Let u be an idempotent in the Ellis semigroup S.

(a) Every ideal in S contains a minimal idempotent.

(b) There exists a minimal idempotent v such that $u >_R v$ and $u >_L v$, i.e. such that $uv = vu = v$.
(c) Every closed co-ideal in S contains a maximal idempotent.
(d) There exists a maximal idempotent v such that $v >_L u$, i.e. $vu = u$.

PROOF. (a), (b): If p is an element of an ideal J then J contains the closed ideal Sp which contains a minimal ideal by the earlier Zorn's Lemma argument. Such a minimal ideal contains idempotents by the Ellis-Numakura Lemma. Apply this to $J = Su$ and we get a minimal idempotent $\tilde{v}$ contained in Su. Let $v = u\tilde{v}$ and apply Lemma 2.2(d).

(c),(d): By (2.11) maximality of an idempotent v corresponds with minimality of the closed co-ideal Iso_v. By Zorn's Lemma the closed co-ideal Iso_u contains a minimal closed co-ideal H. By the Ellis-Numakura Lemma, H contains some idempotent v and since H is a co-ideal $Iso_v \subset H \subset Iso_u$. Thus, v is an idempotent with $v >_L u$. If $w >_L v$ then $Iso_w \subset Iso_v \subset H$ and so, by minimality of H, $Iso_w = H = Iso_v$. Hence, $v >_L w$. Thus, v is a maximal idempotent. For (d) we apply this to $H = Iso_u$. □

Proposition 2.4. Let $\Phi : S \times X \to X$ be an Ellis action. For points $x, x_1, x_2 \in X$

(a) $x \in RECUR$ iff there exists an idempotent $u \in S$ such that $ux = x$ and we can then choose u to be a maximal idempotent.
(b) x is a minimal point of X iff there exists a minimal idempotent $u \in S$ such that $ux = x$. In that case, Iso_x is a closed co-ideal which meets every ideal in S.
(c) $(x_1, x_2) \in PROX$ iff there exists an idempotent $u \in S$ such that $ux_1 = ux_2$ and we can then choose u to be a minimal idempotent. If, in addition, x_1 is a minimal point - a fortiori if X is minimal- then u can be chosen minimal and so that $ux_1 = ux_2 = x_1$.
(d) $(x_1, x_2) \in PROX \wedge RECUR$ iff there exist idempotents $u, v \in S$ such that

(2.12) $$vu = uv = v.$$

(2.13) $$ux_1 = x_1 \text{ and } ux_2 = x_2.$$

(2.14) $$vx_1 = vx_2.$$

and we can then choose v to be minimal. If, in addition, x_1 is a minimal point - a fortiori if X is minimal- then v can be chosen minimal and so that

(2.15) $$vx_1 = vx_2 = x_1.$$

PROOF. In (a),(c),(d) sufficiency is obvious. If u is minimal and $ux = x$ for $x \in X$ then the orbit Sx is image under the continuous action map Φ_x of the minimal ideal Su and so it is minimal.

(a): If $x \in RECUR$ then $H = Iso_x$ is a closed co-ideal and so contains a maximal idempotent u by Proposition 2.3(c).

(b): If J is a minimal ideal then $\Phi_x(J) \subset Sx$ with equality if the latter is minimal. So in that case, $J \cap Iso_x$ contains a, necessarily minimal, idempotent. The closed co-ideal then meets every ideal by Lemma 2.2(e).

(c): If $(x_1, x_2) \in PROX$ then $H = Foc_{(x_1,x_2)}$ is a closed ideal and so it contains a minimal ideal J which contains idempotents. If, in addition x_1 is a minimal point then by part (b) $J \cap Iso_{x_1}$ is a closed subsemigroup and we can choose u in this intersection.

(d): As in (c), let H be the closed ideal $Foc_{(x_1,x_2)}$. From (a) we can choose $u \in Id(S)$ so that (2.13) holds and it implies that $Hu \subset H$. The closed ideal Hu contains a minimal ideal J which contains an idempotent $\tilde{v}$ Since $Hu \subset H$, $\tilde{v} \in H$. Let $v = u\tilde{v}$. Because H is an ideal $v \in H$ and so (2.14) holds. By Lemma 2.2c v is an idempotent which satisfies (2.12).

If, in addition, x_1 is a minimal point then, applying part (b) as in (c), we can choose $\tilde{v}$ in $J \cap Iso_{x_1}$. Proceed as before. □

REMARK 2.1. If u is any idempotent of S and x is any point of X then $uux = ux$ and so ux is recurrent. Thus, from part (a) we obtain the identity:

$$(2.16) \qquad Id(S)X \quad = \quad RECUR$$

It is a useful old theorem of Auslander and Ellis that any point $x \in X$ is proximal to some minimal point, e.g. use ux with u any minimal idempotent. For a recurrent point we can use the above results to sharpen the conclusion.

Theorem 2.5. Let $\Phi : S \times X \to X$ be an Ellis action. If x is a recurrent point of X then there exists a minimal point y which is proximal to x and such that the pair (x, y) is recurrent. That is,

$$(2.17) \qquad y \in Min(X) \quad \text{and} \quad (x,y) \in PROX \wedge RECUR.$$

PROOF. By Proposition 2.4(a) there exists an idempotent u such that $ux = x$. Let $\tilde{v}$ be a minimal idempotent in the ideal Su and let v be the minimal idempotent $u\tilde{v}$ (see Lemma 2.2(d)). Let $y = vx$. By Proposition 2.4(b) y is minimal and by Proposition 2.4(d) $(x, y) \in PROX \wedge RECUR$. □

Lemma 2.6. Let $\Phi : S \times X \to X$ and $\Psi : S \times Y \to Y$ be Ellis actions and $\pi : X \to Y$ be a surjective, continuous action map.

(2.18) $\qquad \pi(RECUR(X)) \quad = \quad RECUR(Y)$

and if Y is minimal then

(2.19)
$$(\pi \times \pi)(PROX(X)) \quad = \quad PROX(Y).$$
$$(\pi \times \pi)(PROX \wedge RECUR(X)) \quad = \quad PROX \wedge RECUR(Y).$$

PROOF. We must prove the inclusions which reverse those of Lemma 1.6(a).

If $y \in RECUR(Y)$ then there exists an idempotent u such that $uy = y$ and there exists $x_1 \in X$ such that $\pi(x_1) = y$. With $x = ux_1$, $\pi(x) = y$ and $ux = x$.

If $(y_1, y_2) = (\pi(x_1), \pi(x_2)) \in PROX(X)$ and Y is minimal then by Proposition 2.4(d) there exists a minimal idempotent u such that $\pi(x_1) = u\pi(x_1) = u\pi(x_2)$. Then $(ux_2, x_2) \in (\pi \times \pi)^{-1}((y_1, y_2)) \cap PROX(X)$.

If $(y_1, y_2) = (\pi(x_1), \pi(x_2)) \in PROX \wedge RECUR(Y)$ and Y minimal then there exist idempotents u, v with v minimal such that $uv = vu = v$, $\pi(x_1) = u\pi(x_1), \pi(x_2) = u\pi(x_2)$, and $\pi(x_1) = v\pi(x_1) = v\pi(x_2)$. $z = (vx_2, ux_2) \in (\pi \times \pi)^{-1}((y_1, y_2))$. Since $uv = v$, $z \in RECUR^2$ and since $vu = v$, $z \in PROX$. That is, $z \in PROX \wedge RECUR(X)$. $\square$

Recall that for any semigroup action $\Phi : S \times X \to X$ we defined the set of transitive points

(2.20) $\qquad TRANS \quad =_{def} \quad \{x \in X : Sx = X\}.$

The system is called *point transitive* when $TRANS$ is nonempty. Clearly, $TRANS \subset RECUR$ and X is minimal iff $TRANS = X$. The system is called *point minimal* when every point is minimal, i.e. $Min(X) = X$. The system is called *totally recurrent* when every point is recurrent, i.e. $RECUR = X$. It is called *orbit minimal* when Sx is minimal for every $x \in X$, and *recurrent point minimal* when every recurrent point is minimal. Clearly,

(2.21)
minimal $\implies$ point minimal $\implies$ orbit minimal $\implies$ recurrent point minimal.

minimal $\iff$ recurrent point minimal and point transitive.

point minimal $\iff$ recurrent point minimal and totally recurrent.

When the action is extended to S^+ the concept of minimal set does not change. In particular, $J \subset S$ is a minimal ideal in S iff it is a minimal ideal in S^+. On the other hand, since every point is recurrent for the S^+ action, it follows that for the S^+ action X is point minimal iff it is recurrent point minimal and so iff it is orbit minimal.

It can be argued that the versatility of the distality concept (and its relatives) grows out of the number of apparently unrelated conditions which characterize distality. We assemble these now in a fashion which highlights the parallelism between distality, almost distality and semidistality.

Theorem 2.7. Let $\Phi : S \times X \to X$ be an Ellis action.
  (a) The following properties are equivalent
      (i) Φ is distal.
      (ii) The product system Φ^2 on $X \times X$ is point minimal.
      (iii) The enveloping semigroup $\Phi^\#(S)$ is minimal and contains the identity map 1_X.
      (iv) The idempotents of $\Phi^\#(S)$ are injective maps on X.
      (v) The idempotents of $\Phi^\#(S)$ are surjective maps on X.
      (vi) The identity map 1_X is contained in $\Phi^\#(S)$ and is the unique idempotent in $\Phi^\#(S)$.
      (vii) $\Phi^\#(S)$ is a group of bijections on X.
      (viii) Some minimal element of $\Phi^\#(S)$ is an injective map.
      (ix) Some minimal element of $\Phi^\#(S)$ is a surjective map.
      If Φ is distal then it is point minimal.
  (b) The following properties are equivalent
      (i) Φ is almost distal.
      (ii) The product system Φ^2 on $X \times X$ is orbit minimal.
      (iii) The enveloping semigroup $\Phi^\#(S)$ is minimal.
      If Φ is almost distal then it is orbit minimal.
  (c) The following properties are equivalent
      (i) Φ is semidistal.
      (ii) The product system Φ^2 on $X \times X$ is recurrent point minimal.
      (iii) Every idempotent of the enveloping semigroup $\Phi^\#(S)$ is minimal in $\Phi^\#(S)$.
      If Φ is semidistal then it is recurrent point minimal.

PROOF. We will prove (b) and (c) first and then use these results for the, more elaborate, (a). Recall that the three concepts distality, almost distality and semidistality agree for the S^+ action and that they coincide with distality for S. Similarly, point minimality, orbit minimality and recurrent point minimality agree for the S^+ action and they coincide with point minimality for S.

(b), (iii) $\Rightarrow$ (i): Let $(x_1, x_2) \in PROX$. By Proposition 2.4(c) there is a minimal idempotent $u \in Foc_{(x_1,x_2)}$. The image $\Phi^\#(Su)$ of the minimal ideal Su is equal, by assumption (iii), to $\Phi^\#(S)$. Since $Foc_{(x_1,x_2)}$ is an ideal $Su \subset Foc_{(x_1,x_2)}$ and so (1.5) implies $p \in Foc_{(x_1,x_2)}$ for all $p \in S$. Hence, $(x_1, x_2) \in ASYMP$.

(i) $\Rightarrow$ (ii): For $x \in X$ and u any minimal idempotent in S, the pair (x, ux) is proximal and so is asymptotic for an almost distal system. Thus, $px = pux$ for every $p \in S$ and so the orbit Sx is the image under the continuous action map Φ_x of the minimal ideal Su. Hence, it is a minimal invariant set. Thus, any almost distal system is orbit minimal. By Proposition 1.9 the product system Φ^2 is almost distal when Φ is and so $X \times X$ is orbit minimal.

(ii) $\Rightarrow$ (iii): Let u be any minimal idempotent in S. For all $x \in X$ $(x, ux) \in PROX$ and so the orbit $S(x, ux)$, which is minimal by assumption (ii), meets, and so is contained in, the diagonal, 1_X. Hence, for all $p \in S$, $px = pux$. Since this is true for all x, $\Phi^\#(p) = \Phi^\#(pu)$ in X^X. Thus, the enveloping semigroup $\Phi^\#(S)$ is the image of every minimal ideal Su in S and so is minimal.

(c), (iii) $\Rightarrow$ (i): Assume that every idempotent in $\Phi^\#(S)$ is minimal. Since $PROX \wedge RECUR_S = PROX \wedge RECUR_{\Phi^\#(S)}$ we can replace S by the enveloping semigroup $\Phi^\#(S)$ and so assume that every idempotent in S is minimal. Let $(x_1, x_2) \in PROX \wedge RECUR$. By Lemma 2.3(d) there exist idempotents $u, v \in A$ which satisfy equations (2.12), (2.13), (2.14). Since $vu = v$, $v \in Su$. By assumption u is a minimal idempotent and so $Sv = Su$ which implies $uv = u$. Then equation (2.12) implies $u = v$ and so (2.13) and (2.14) imply that $x_1 = x_2$.

(i) $\Rightarrow$ (ii): If x is a recurrent point then there exists an idempotent u such that $ux = x$. By Lemma 2.3(d) there is a minimal idempotent v such that $uv = vu = v$. The pair $(x, vx) \in PROX \wedge RECUR$ and so by assumption (i), $x = vx$. Thus, any semidistal system is recurrent point minimal. By Proposition 1.9 the product system Φ^2 is semidistal when Φ is and so Φ semidistal implies $X \times X$ is recurrent point minimal.

(ii) $\Rightarrow$ (iii): Let u be any idempotent in S. As before, we can obtain from Lemma 2.3(d) a minimal idempotent $v \in Su$ such that $uv = vu = v$. So for any $x \in X$ the pair $(x_1, x_2) = (ux, vx)$ is in $PROX \wedge RECUR$. Because the pair is proximal, its orbit meets the diagonal. Because the pair is recurrent, assumption (ii) implies that its orbit is contained in the diagonal. Hence, $ux = vx$ for all $x \in X$. That is, $\Phi^\#(u) = \Phi^\#(v)$ and so $\Phi^\#(u)$ is a minimal idempotent in $\Phi^\#(S)$.

(a): As promised above, we apply (b) to the S^+ extension of the S action. $\Phi^\#(S)$ is a closed S^+ ideal in $\Phi^\#(S^+) = \{1_X\} \cup \Phi^\#(S)$ and so $\Phi^\#(S^+)$ is minimal iff $\Phi^\#(S)$ is minimal and, in addition, $1_X \in \Phi^\#(S)$. Thus, (i), (ii) and (iii) are equivalent and, in addition, distality implies point minimality.

(vii) $\Rightarrow$ (iv) and (v): Obvious.

(iv) or (v) $\Rightarrow$ (vi): Any idempotent $u \in X^X$ agrees with the identity on the image set $u(X)$ and maps the pair (x, ux) to the pair (ux, ux). Hence, if u is either injective or surjective then $u = 1_X$. Thus, either (iv) or (v) implies that 1_X is the only idempotent in $\Phi^\#(S)$.

(vi) $\Rightarrow$ (iii): If $1_X \in \Phi^\#(S)$ and 1_X is a minimal idempotent in $\Phi^\#(S)$ then $\Phi^\#(S) = \Phi^\#(S)1_X$ is minimal.

(iii) ⇒ (vii): By assumption (iii) 1_X is a minimal idempotent in $\Phi^\#(S)$ and $\Phi^\#(S)$ is minimal. For all $p \in \Phi^\#(S)$ there exists $q \in \Phi^\#(S)$ such that $qp = 1_X$. As every map in the monoid $\Phi^\#(S)$ has a left inverse (vii) follows.

(iii) ⇒ (viii) and (ix): Obvious.

(viii) ⇒ (iii): For minimal element $p \in \Phi^\#(S)$ the ideal $J = \Phi^\#(S)p$ contains an idempotent u. Since a minimal element is recurrent, $p \in J$. Since J is a minimal ideal, $Ju = J$. Hence, $pu = p$. From assumption (viii) we can begin with p an injective map. $pu = p$ implies that the minimal idempotent u is injective. As mentioned above, 1_X is the only injective idempotent in X^X. Since $u = 1_X$ is a minimal idempotent $\Phi^\#(S) = \Phi^\#(S)1_X$ is minimal and so (iii) holds.

(ix) ⇒ (iii): For minimal element $p \in \Phi^\#(S)$ the ideal $J = \Phi^\#(S)p$ contains p and satisfies $Jp = J$. So $Iso_p \cap J$ is a nonempty closed semigroup which contains an idempotent v. $vp = p$ and v is a minimal idempotent. From assumption (ix) we can begin with p a surjective map. $vp = p$ implies that the minimal idempotent v is surjective. As mentioned above, 1_X is the only surjective idempotent in X^X. Since $v = 1_X$ is a minimal idempotent $\Phi^\#(S) = \Phi^\#(S)1_X$ is minimal and so (iii) holds. □

REMARK 2.2. Part (a) is the Ellis characterization of distality. The characterization of almost distality in part (b) is due to Blanchard et al. (2002). Part (c) continues the analogy to semidistality.

Corollary 2.8. *A point transitive, semidistal system is minimal.*

PROOF. If a system contains a minimal, transitive point then it is minimal. Hence, a point transitive system which is recurrent point minimal is minimal. The result then follows from part (c). □

Theorem 2.9. *Let $\Phi : S \times X \to X$ and $\Psi : S \times Y \to Y$ be Ellis actions and $\pi : X \to Y$ be a surjective, continuous action map.*
  (a) *If X is point transitive, minimal, point minimal, orbit minimal or recurrent point minimal then Y satisfies the corresponding property.*
  (b) *If the action Φ is proximal, asymptotic, distal, almost distal or semidistal then both the action Ψ and the map π satisfy the corresponding property.*
  (c) *If both the action Ψ and the map π are either proximal, distal or semidistal then the action Φ satisfies the corresponding property.*
  (d) *If the action Ψ is distal and the map π is asymptotic then the action Φ is almost distal.*

(e) Assume the map π is proximal. The map π is almost distal if and only if it is asymptotic. In particular, if Φ is almost distal then π is asymptotic. The map π is semidistal if and only if

(2.22) $$RECUR^2 \cap R_\pi \subset 1_X.$$

In particular, the latter condition holds if Φ is semidistal.

PROOF. (a): By (1.21) π maps $TRANS(X)$ into $TRANS(Y)$ and so the result for point transitivity follows. Since Φ is minimal exactly when $X = TRANS$ the results for minimality and hence for point minimality and orbit minimality follow as well. By (2.18) every recurrent point in Y is the image of one in X and so we get the result for recurrent point minimality.

(b): Applying the results of part (a) to $\pi \times \pi$ we see from Theorem 2.7 that the three variants of distality are inherited by factors as well. By Proposition 1.7(b), with ϵ the trivial action map, proximality and asymptoticity are inherited by factors. That the properties for Φ imply the corresponding properties for π follow from Proposition 1.7(a) with ϵ again the trivial map.

(c): Apply Proposition 1.7(d) with ϵ trivial.

(d): If $z \in PROX(X)$ then $w = (\pi \times \pi)(z) \in PROX(Y)$. Since Ψ is distal $w \in 1_Y$ and so $z \in R_\pi \subset ASYMP(X)$.

(e): By (b) the map π is almost distal (or semidistal) if Φ is almost distal (resp. semidistal). Since π proximal says $R_\pi \subset PROX$ the map property equivalences are obvious. □

In light of Theorem 2.7 it is useful to be able to reduce map properties to space properties, i.e. to obtain the converse of Proposition 1.4. This can be accomplished in some cases when the base action is minimal.

Theorem 2.10. Let $\Phi : S \times X \to X$ and $\Psi : S \times Y \to Y$ be Ellis actions and $\pi : X \to Y$ be a continuous action map. Assume that Y is minimal. The action map π is proximal, distal or semidistal iff for every $y \in Y$ the Iso_y action on the fiber $X_y = \pi^{-1}(y)$, $\Phi(y) : Iso_y \times X_y \to X_y$, satisfies the corresponding property.

PROOF. If a pair $(x_1, x_2) \in X \times X$ is proximal and Y is minimal then we can first choose $p \in S$ so that $px_1 = px_2$ and then $q \in S$ so that $qp\pi(x_1) = \pi(x_1)$. It follows that any pair in $X_y \times X_y$ is proximal iff it is Iso_y proximal. Recall from the proof of Proposition 1.4 itself that any pair in $X_y \times X_y$ is recurrent iff it is Iso_y recurrent. From these observations the proofs consist of three easy exercises. □

Since R_π is a closed invariant subset of $X \times X$ we can can restrict the action of Φ^2 to R_π.

Corollary 2.11. Let $\Phi : S \times X \to X$ and $\Psi : S \times Y \to Y$ be Ellis actions and $\pi : X \to Y$ be a continuous action map. Assume that Y is minimal.
  (a) π is a distal action map iff R_π is point minimal.
  (b) π is a semidistal action map iff R_π is recurrent point minimal.

PROOF. As usual, it suffices to prove (b) and then obtain (a) by using the extended S^+ actions.

By Theorem 2.10 and Theorem 2.7(c), π is semidistal iff for every $y \in Y$ the product space $X_y \times X_y$ is recurrent point minimal with respect to the Iso_y action. Let $z \in X_y \times X_y$. If $p \in S$ and $pz = z$ then $p \in Iso_y$. That is, $z \in R_\pi$ is recurrent with respect to S iff $z \in X_y \times X_y$ is recurrent with respect to Iso_y. By Proposition 2.4(b) Iso_y meets every minimal ideal of S because Y is minimal. Thus, Lemma 2.2(e) implies that p is minimal in S iff it is minimal in Iso_y. Hence, from Proposition 2.4(b) z is a minimal point of R_π for the S action iff it is a minimal point of $X_y \times X_y$ for the Iso_y action. $\square$

The following is a map version of the factor results of Theorem 2.9(b).

Proposition 2.12. Let $\Phi : S \times X \to X$, $\Psi : S \times Y \to Y$ and $\Theta : S \times Z \to Z$ be Ellis actions and $X \xrightarrow{\pi} Y \xrightarrow{\epsilon} Z$ be continuous action maps with π surjective.
  (a) Assume that Z is minimal. If the composition $\epsilon \circ \pi$ is a distal or semidistal map then ϵ satisfies the corresponding property.
  (b) Assume that Y is minimal. If the composition $\epsilon \circ \pi$ is an almost distal map then ϵ is almost distal.

PROOF. (a): It suffices by Theorem 2.10 to show that for all $z \in Z$ the Iso_z action on Y_z is distal or semidistal. By Proposition 1.4 the Iso_z action on X_z satisfies the required property and so the result follows by applying Theorem 2.9(b) to the Iso_z action map $\pi : X_z \to Y_z$. That is, we have reduced the factor result for maps to the corresponding one for spaces.

(b): Here we must proceed directly. Given $(y_1, y_2) \in R_\epsilon \cap PROX(Y)$ then because Y is minimal, Lemma 2.6 implies that there exists a pair $(x_1, x_2) \in PROX(X)$ which maps onto (y_1, y_2). Since $\epsilon \circ \pi$ is almost distal and $(x_1, x_2) \in R_{\epsilon \circ \pi}$ it follows that (x_1, x_2) is asymptotic and so (y_1, y_2) is as well. $\square$

REMARK 2.3. Observe that if Y is minimal then Z is and so the assumption required in (b) is stronger than that in (a).

A surjective, continuous action map $\pi : X \to Y$ is called *minimal* if the only closed invariant subset K of X such that $\pi(K) = Y$ is X itself. Clearly, X is minimal iff the trivial action map on X is minimal.

Proposition 2.13. Let $\Phi : S \times X \to X$, $\Psi : S \times Y \to Y$ and $\Theta : S \times Z \to Z$ be Ellis actions and $X \xrightarrow{\pi} Y \xrightarrow{\epsilon} Z$ be surjective, continuous action maps.
  (a) The composed action map $\epsilon \circ \pi$ is a minimal action map iff both ϵ and π are minimal action maps.
  (b) The space X is minimal iff the action map π is minimal and the space Y is minimal.
  (c) If the space Y is point transitive then the following are equivalent.
    (i) π is minimal.
    (ii) $TRANS(\Phi) = (\pi)^{-1}(TRANS(\Psi))$.
    (iii) There exists $y \in TRANS(\Psi)$ such that $(\pi)^{-1}(y) \subset TRANS(\Phi)$.
  When these conditions hold X is point transitive.

PROOF. (a): Assume π and ϵ are minimal and $K \subset X$ is closed and invariant with $\epsilon(\pi(K)) = Z$. Since $\pi(K)$ is closed and invariant and ϵ is minimal $\pi(K) = Y$. Since π is minimal $K = X$. Assume that $\epsilon \circ \pi$ is minimal and $K \subset X$ is closed and invariant with $\pi(K) = Y$. Then $\epsilon(\pi(K)) = Z$ and so $K = X$ because $\epsilon \circ \pi$ is minimal. If $L \subset Y$ and $\epsilon(L) = Z$ then $K = (\pi)^{-1}(L)$ satisfies $\epsilon(\pi(K)) = Z$ and so $K = X$. Hence, $L = \pi(K) = Y$.

(b): Apply (a) with ϵ the trivial action map.

(c), (i) $\Rightarrow$ (ii): If $\pi(x) \in TRANS(\Psi)$ then $K = Sx$ is a closed invariant set which maps onto Y. By assumption (i) $X = Sx$ and so $x \in TRANS(\Phi)$. This proves $TRANS(\Phi) \supset (\pi)^{-1}(TRANS(\Psi))$. The reverse inclusion follows from Lemma 1.6(a). Since $TRANS(\Phi)$ is nonempty, X is transitive.

(ii) $\Rightarrow$ (iii): Obvious.

(iii) $\Rightarrow$ (i): Let y satisfy the condition of (iii). If $K \subset X$ be invariant with $\pi(K) = Y$. There exists $x \in K$ with $\pi(x) = y$ and so assumption (iii) implies $x \in TRANS(\Phi)$. Since K is invariant $X = Sx \subset K$. □

Theorem 2.14. Let $\Phi : S \times X \to X$ and $\Psi : S \times Y \to Y$ be Ellis actions and $\pi : X \to Y$ be a surjective continuous action map. If X is point transitive and π is semidistal then Y is point transitive and π is minimal. If, in addition, Y is minimal then X is minimal.

PROOF. Let $x \in TRANS(X)$ so that $y = \pi(x) \in TRANS(Y)$ by (1.21). In particular, Y is point transitive. Now let K be a closed invariant subset of X with $\pi(K) = Y$. To show that π is minimal it is sufficient to show that $x \in K$. Since $y \in Y$ is recurrent, we can consider the Iso_y action on the fiber X_y. Notice that $px \in X_y$ implies $p \in Iso_y$ and so $x \in X_y$ is a transitive point for the Iso_y action. Since π is assumed to be semidistal, Proposition 1.4 implies that the Iso_y action on X_y is semidistal and so by Corollary 2.8, it is minimal. Since $\pi(K) = Y$, $K \cap X_y$ is nonempty and so is a closed, Iso_y invariant subset of the Iso_y minimal space X_y. Hence, $X_y \subset K$ as required.

If Y is minimal then X is minimal by Proposition 2.13(b). □

Corollary 2.15. Let $\Phi : S \times X \to X$ and $\Psi : S \times Y \to Y$ be Ellis actions and $\pi : X \to Y$ be a surjective continuous action map. If Y is point minimal and π is distal then X is point minimal. If Y is recurrent point minimal and π is semidistal then X is recurrent point minimal.

PROOF. Assume that π is semidistal and that x is a recurrent point in X with $y = \pi(x)$. The restricted map $\pi|Sx : Sx \to Sy$ is semidistal map of S actions. Since x is a recurrent point, it lies in its orbit Sx and is a transitive point for Sx. If Y is recurrent point minimal then Sy is minimal and so by Theorem 2.14 Sx is minimal. Thus, X is recurrent point minimal.

To obtain the first result we apply the second to the extended S^+ actions in the now familiar way. □

For an Ellis action $\Phi : S \times X \to X$, $[RECUR(X)]$ denotes the smallest closed invariant subset of X which contains the recurrent points of X. We call this the *center* of X. Similarly, the *min-center*, $[Min(X)]$, is the smallest closed invariant subset of X which contains the minimal points of X.

Proposition 2.16. Let $\Phi : S \times X \to X$ and $\Psi : S \times Y \to Y$ be Ellis actions and $\pi : X \to Y$ be a surjective continuous action map. If π is an asymptotic map and $X = [RECUR]$ then the map π is minimal.

PROOF. Let K be a closed invariant subset of X with $\pi(K) = Y$. To show that π is minimal it is sufficient to show that $RECUR \subset K$. If $x \in RECUR$ then there exists $p \in S$ such that $px = x$. Since π is surjective there exists $x_1 \in K$ with $\pi(x) = \pi(x_1)$. Because π is asymptotic, $x = px = px_1$. Since K is invariant, it follows that $x \in K$. □

Distality and its relatives are conditions upon the entire system. We end this section by describing the results about the related pointwise condition from Auslander and Furstenberg (1994). Let $\Phi : S \times X \to X$ be an Ellis action and $x \in X$. We call x a *distal point* of X if

(2.23) $\qquad x' \in Sx \quad \text{and} \quad (x, x') \in PROX(X) \quad \implies \quad x = x'.$

We call x a *product recurrent point* if for every Ellis action $\Psi : S \times Y \to Y$

(2.24) $\quad y \in RECUR(Y) \quad \Longrightarrow \quad (x,y) \in RECUR(X \times Y)$.

Theorem 2.17. Let $\Phi : S \times X \to X$ be an Ellis action and $x \in X$.
  (a) The following conditions on the point x are equivalent.
      (i) x is product recurrent.
      (ii) Iso_x contains every maximal idempotent of S.
      (iii) $wx = x$ for every maximal idempotent w of S.
      (iv) For every maximal idempotent w, the pair (x, w) is a recurrent point in $X \times S$.
      (v) $p \in RECUR(S)$ implies $(x, p) \in RECUR(X \times S)$.
  (b) The following conditions on the point x are equivalent.
      (i) x is a distal point.
      (ii) Iso_x contains every idempotent of S.
      (iii) $ux = x$ for every idempotent u of S.
      (iv) $ux = x$ for every minimal idempotent u of S.
      (v) There exists an ideal J of S such that $ux = x$ for every minimal idempotent u of J.
      (vi) $x' \in Min(X)$ implies $(x, x') \in Min(X \times X)$.
      (vii) There exists a minimal ideal J of S such that $(x, p) \in Min(X \times S)$ for all $p \in J$.
      (viii) For every Ellis action $\Psi : S \times Y \to Y$

(2.25) $\quad y \in Min(Y) \quad \Longrightarrow \quad (x, y) \in Min(X \times Y)$.

  (c) A distal point is minimal and product recurrent.
  (d) The system Φ is distal iff every point of X is a distal point.

PROOF. (a), (i) $\Rightarrow$ (v): Obvious.
(v) $\Rightarrow$ (iv): Every idempotent is a recurrent point.
(iv) $\Rightarrow$ (iii): By Proposition 2.4(a) there exists an idempotent u such that $ux = x$ and $uw = w$. In particular, $u >_L w$. By maximality of w it follows that $wu = u$. Hence, $wx = wux = ux = x$.
(iii) $\Rightarrow$ (ii): Obvious.
(ii) $\Rightarrow$ (i) : If y is a recurrent point then by Proposition 2.4(a) there exists a maximal idempotent w such that $wy = y$. Since $w \in Iso_x$ we have $w(x, y) = (x, y)$ and so the pair is recurrent.
(b), (i) $\Rightarrow$ (ii): If u is an idempotent then $ux \in Sx$ and $(x, ux) \in PROX$. Hence, if x is a distal point, $ux = x$.
(ii) $\Rightarrow$ (iii), (iii) $\Rightarrow$ (iv), (iv) $\Rightarrow$ (v): Obvious.
(v) $\Rightarrow$ (viii): Any ideal contains a minimal ideal and so we can assume that J in (v) is minimal. If y is a minimal point of Y then $y \in Sy = Jy$ and so $Iso_y \cap J$ contains an idempotent u which is minimal because J is. By assumption (v) $ux = x$ and so $u(x, y) = (x, y)$. It follows from Proposition 2.4(b) that (x, y) is a minimal point.

(viii) ⇒ (vii): Obvious.

(vii) ⇒ (vi): $1_X \times \Phi_{x'} : X \times J \to X \times X$ is an action map and the image of the set of minimal points $\{x\} \times J$ contains (x, x').

(vi) ⇒ (i): Since projection to the first coordinate is an action map, (x, x') minimal implies x is minimal. Since X contains some minimal point x' it follows from condition (vi) that x is a minimal point. Hence, if x' is any point of Sx then x' is minimal and so by (vi) (x, x') is minimal. If also (x, x') is proximal then $S(x, x')$ meets the diagonal 1_X and so is contained in the diagonal by minimality. A minimal point is recurrent and so $(x, x') \in S(x, x') \subset 1_X$. Thus, $x = x'$ as required.

(c): We have just shown that a distal point is necessarily minimal. Comparing (ii) of (a) and (b), we see that a distal point is product recurrent.

(d): Clearly, if Φ is distal then every point is a distal point. On the other hand, if every point is a distal point then every idempotent of S acts as the identity map by (b) (iii). It follows from Theorem 2.7(a) that the system is distal. □

Corollary 2.18. Let $\Phi : S \times X \to X$ and $\Psi : S \times Y \to Y$ be Ellis actions and $\pi : X \to Y$ be a surjective continuous action map. Let $x \in X$.
- (a) If x is a distal point (or a product recurrent point) of X then $\pi(x)$ is a distal point (resp. a product recurrent point) of Y.
- (b) Assume that $\{x\} = \pi^{-1}(\{\pi(x)\})$. If $\pi(x)$ is a distal point (or a product recurrent point) of Y then x is a distal point (resp. a product recurrent point) of X.
- (c) Assume X is minimal. If x is a distal point of X and π is a proximal map then $\{x\} = \pi^{-1}(\{\pi(x)\})$.

PROOF. (a),(b): If $ux = x$ then $u\pi(x) = \pi(x)$. If $u\pi(x) = \pi(x)$ and $\{x\} = \pi^{-1}(\{\pi(x)\})$ then $ux = x$. Apply condition (iii) of (a) and (b) in Theorem 2.17.

(c): If π is a proximal map and $\pi(x') = \pi(x)$ then $(x, x') \in PROX$. Since X is minimal, $X = Sx$ and and so $x' \in Sx$. If, in addition, x is a distal point, then $x = x'$. □

If S is a monoid then every point is recurrent and so every point is product recurrent. Thus, a product recurrent point need not even be minimal in general. However, as we will later see, in certain classical systems product recurrence implies distality.

CHAPTER 3

Continuity Conditions

The intuitions developed for a classical dynamical system (X, f) can be misleading when applied to Ellis actions. For example, let $\Phi : S \times X \to X$ be an Ellis action and suppose that for every $p \in S$ the map Φ^p is surjective. In particular, every idempotent in the enveloping semigroup $\Phi^\#(S)$ is surjective. This implies that the action is distal. In fact, the following is just a restatement of part of Theorem 2.7(a).

Theorem 3.1. Let $\Phi : S \times X \to X$ be an Ellis action. The action is distal iff the map $\Phi^p : X \to X$ is surjective for every $p \in S$. In that case, the enveloping semigroup $\Phi^\#(S)$ is a group of bijections on X.

If f is a homeomorphism on X then of course all the iterates $\{f^n : n \in \mathbb{Z}\}$ are continuous and surjective. However, for the associated Ellis action of $S = \beta\mathbb{Z}_+$ or $\beta\mathbb{Z}$ on X, the maps Φ^p are usually neither surjective nor continuous. So, of course, the action map $\Phi : S \times X \to X$ is usually not continuous.

We now consider the meaning of the assumption of joint continuity for the multiplication $M : S \times S \to S$ and for the action $\Phi : S \times X \to X$.

Definition 3.2. An Ellis semigroup S is a *topological semigroup* when the multiplication $M : S \times S \to S$ is jointly continuous. An Ellis action Φ is called a *topological action* when $\Phi : S \times X \to X$ is jointly continuous.

N.B. As we will see, an Ellis action can be a topological action even when S is not a topological semigroup. Of course, S is a topological semigroup exactly when the translation action is a topological action.

Proposition 3.3. (a) Let S be an Ellis semigroup which is a group. If S is a topological semigroup then it is a topological group. That is, the inversion map $inv : S \to S$ is continuous.
(b) Let $g : S \to T$ be a surjective, continuous homomorphism of Ellis semigroups. If S is a topological semigroup then T is a topological semigroup.
(c) Let $\Phi : S \times X \to X$ and $\Psi : S \times Y \to Y$ be Ellis actions and let $\pi : X \to Y$ be a surjective, continuous action map. If Φ is a topological action then Ψ is a topological action.

(d) Let $\Phi : S \times X \to X$ be an Ellis action. If S is a topological semigroup then the enveloping semigroup $\Phi^\#(S)$ is a topological semigroup.

(e) An Ellis action $\Phi : S \times X \to X$ is a topological action iff the Ellis action $Ev : \Phi^\#(S) \times X \to X$ is a topological action.

(f) Let $\Phi : S \times X \to X$ be an Ellis action and $K \subset X$ be closed and invariant. If Φ is a topological action then the restriction $\Phi|K : S \times K \to K$ is a topological action.

PROOF. (a): A map on a compact Hausdorff space is continuous iff it has a closed graph. $inv = M^{-1}(\{e\})$ which is closed when M is continuous.

(b): Because g is a homomorphism, $M_T \circ (g \times g) = g \circ M_S$ which is continuous when S is a topological semigroup. Because a continuous surjection of compacta is a quotient map it follows that M_T is continuous.

(c): Because π is an action map, $\Psi \circ (1_S \times \pi) = \pi \circ \Phi$. Proceed as in (b).

(d): $\Phi^\# : S \to \Phi^\#(S)$ is a continuous, surjective homomorphism. Apply (b).

(e): $Ev \circ (\Phi^\# \times 1_X) = \Phi$. So Φ is continuous if Ev is. As before the converse result follows because $\Phi^\# \times 1_X$ is a quotient map.

(f): Obvious. $\square$

We will need a brief review of some standard function space facts. Since our spaces X are compact Hausdorff the results are easy to summarize.

Let $\mathcal{C}(X, X)$ denote the semigroup of continuous automorphisms of X, equipped with the *compact-open topology*. That is, we use as subbase the collection of sets

(3.1) $$\mathcal{J}(A, U) \quad =_{def} \quad \{g \in \mathcal{C}(X, X) : g(A) \subset U\}$$

for $A, U \subset X$ with A compact and U open. Let $\mathcal{C}_s(X, X)$ denote the subspace of continuous surjections on X and let $\mathcal{H}(X)$ denote the subspace of bijections on X, i.e. the homeomorphisms on X.

Clearly, the inclusion map $\mathcal{C}(X, X) \to X^X$ is continuous. That is, the pointwise topology is coarser than the compact-open topology.

A subset $F \subset X^X$ is called *equicontinuous* if for every neighborhood V of the diagonal $1_X \subset X \times X$ there exists a neighborhood W of the diagonal such that $FW \subset V$. That is, $(px_1, px_2) \in V$ for all $p \in F$ and $(x_1, x_2) \in W$. It suffices to show this for closed neighborhoods V which shows that the closure in X^X, i.e. the pointwise closure, of an equicontinuous subset is equicontinuous. Any subset of an equicontinuous set is equicontinuous and a finite set is equicontinuous iff each member is continuous. Of course, any equicontinuous subset is contained in $\mathcal{C}(X, X)$.

We now collect the results we will need.

Proposition 3.4. Let X be a nonempty, compact, Hausdorff space.

(a) $\mathcal{C}(X,X)$ is a, usually noncompact, topological semigroup and the evaluation action of $\mathcal{C}(X,X)$ on X is a topological action. That is, the maps:

(3.2) $$Comp : \mathcal{C}(X,X) \times \mathcal{C}(X,X) \to \mathcal{C}(X,X) \quad Comp(f,g) = f \circ g$$
$$Ev : \mathcal{C}(X,X) \times X \to X \quad Ev(f,x) = f(x)$$

are continuous maps. Furthermore,

(3.3) $$Inv : \mathcal{H}(X) \to \mathcal{H}(X) \quad Inv(f) = f^{-1}$$

is continuous and so $\mathcal{H}(X)$ is a topological group.

(b) $\mathcal{C}_s(X,X)$ is a closed subset of $\mathcal{C}(X,X)$.

(c) (Arzela-Ascoli) A subset $F \subset X^X$ is equicontinuous iff it is a set of continuous functions with compact closure in $\mathcal{C}(X,X)$. In that case, the topologies induced on F from $\mathcal{C}(X,X)$ and from X^X agree.

PROOF. (a): If $f \circ g \in \mathcal{J}(A,U)$ with A compact and U open then we can choose an open set V with closure B such that $g(A) \subset V \subset B \subset f^{-1}(U)$. Then $Comp$ maps $\mathcal{J}(B,U) \times \mathcal{J}(A,V)$ into $\mathcal{J}(A,U)$. The proof for evaluation is similar. See also Kelley (1955) Theorem 7.5. From continuity of $Comp$ it follows that for each $f \in \mathcal{H}(X)$ the map

(3.4) $$Ad_f : \mathcal{C}(X,X) \to \mathcal{C}(X,X) \quad Ad_f(g) = f \circ g \circ f^{-1}$$

is continuous and so is a homeomorphism with inverse $Ad_{f^{-1}}$. Next, check that for $f \in \mathcal{H}(X)$ we can obtain a neighborhood base by letting V vary over neighborhoods of the diagonal 1_X and using

(3.5) $$\mathcal{N}_f(V) =_{def} \{g \in \mathcal{H}(X) : (x, gf^{-1}(x)) \in V \text{ for all } x \in X\}$$
$$= \{g \in \mathcal{H}(X) : (f(x), g(x)) \in V \text{ for all } x \in X\}$$

Continuity of Inv at f is easily checked using such neighborhoods and the continuity of $Ad_{f^{-1}}$.

(b): If f is not surjective then there is a proper open subset U of X such that $f(X) \subset U$. None of the maps in $\mathcal{J}(X,U)$ is surjective. Hence, the complement of $\mathcal{C}_s(X,X)$ is open.

(c): For the classical Arzela-Ascoli Theorem see Kelley (1955) Theorem 7.17. If F is a compact subset of $\mathcal{C}(X,X)$ then the inclusion map to X^X restricts to a homeomorphism of F onto its image. That is, the compact-open and pointwise topologies agree on F and hence on any subset of F. $\square$

Let $\Phi : S \times X \to X$ be an action of a semigroup S on a space X. We will call Φ an *equicontinuous action* when the set of maps $\{\Phi^p : p \in S\}$ is an equicontinuous subset of X^X.

Theorem 3.5. Let $\Phi : S \times X \to X$ be an Ellis action and let D be a dense subset of S. The following conditions are equivalent :

(1) Φ is a topological action, i.e. $\Phi : S \times X \to X$ is jointly continuous.
(2) The action Φ is equicontinuous, i.e. the enveloping semigroup $\Phi^\#(S)$ is an equicontinuous subset of X^X.
(3) The image $\Phi^\#(D)$ is an equicontinuous subset of X^X.
(4) The map $\Phi^p : X \to X$ is continuous for each $p \in S$ and the topologies on $\Phi^\#(S)$ induced by the compact-open and pointwise topologies agree.
(5) The map $\Phi^p : X \to X$ is continuous for each $p \in S$ and the map $\Phi^\# : S \to \mathcal{C}(X, X)$ is continuous.

When these conditions hold the Ellis semigroup $\Phi^\#(S)$ is a topological semigroup. That is, the multiplication map on $\Phi^\#(S)$, which is the restriction of the composition map $Comp : \Phi^\#(S) \times \Phi^\#(S) \to \Phi^\#(S)$, is jointly continuous.

PROOF. (1) $\Rightarrow$ (2): Let V be a neighborhood of the diagonal 1_X in $X \times X$. By continuity of Φ the set

(3.6) $\qquad G = \{(p, x_1, x_2) : (px_1, px_2) \in V\}$

is a neighborhood of $S \times 1_X$ in $S \times X \times X$. By compactness, there exists a neighborhood W of 1_X in $X \times X$ such that $S \times W \subset G$ (see, e.g. Wallace's Theorem 5.12 in Kelley (1955)). Hence, the collection of maps $\{\Phi^p : p \in S\}$ is equicontinuous.

(2) $\Leftrightarrow$ (3). Since D is dense in S and $\Phi^\#$ is continuous, $\Phi^\#(D)$ is dense in the enveloping semigroup with respect to the pointwise topology. As mentioned above the pointwise closure of an equicontinuous set of functions is equicontinuous and so the enveloping semigroup $\Phi^\#(S)$ is equicontinuous when $\Phi^\#(D)$ is. The reverse implication is obvious.

(2) $\Rightarrow$ (4): The two topologies agree by Proposition 3.4(c).

(4) $\Rightarrow$ (5): By assumption (4) the continuous map $\Phi^\# : S \to X^X$ maps onto a set where the pointwise topology agrees with that induced from $\mathcal{C}(X, X)$.

(5) $\Rightarrow$ (1): Clearly, Φ is the composition $Ev \circ (\Phi^\# \times 1_X)$ and so it is continuous by Proposition 3.4(a) and assumption (5).

Condition (4) and Proposition 3.4(a) imply that the composition map is jointly continuous for $\Phi^\#(S)$. $\square$

REMARK 3.1. The identity action of any Ellis semigroup S on a space X with $\Phi^p = 1_X$ for all p, is an equicontinuous action with $\Phi^\#(S)$ the trivial, i.e. singleton, topological group. We cannot conclude that S itself is a topological semigroup.

3. CONTINUITY CONDITIONS

In particular, it follows that equicontinuity for a classical system (X, f), i.e. equicontinuity of the set of iterates $\{f^n : n \in \mathbb{Z}_+\}$, is the same as equicontinuity for the associated Ellis action of $\beta\mathbb{Z}_+$. On the other hand, an equicontinuous homeomorphism is always distal, but for general Ellis actions there exist point transitive, equicontinuous actions which are not distal. For example, let T be any compact group and let $S = T \cup \{z\}$ where z is an isolated point not in T. We extend the multiplication of T by defining:

$$(3.7) \qquad pz = z = zp \qquad \text{for all} \quad p \in S.$$

The translation action of S on itself is clearly equicontinuous and $TRANS = T$ for this action. On the other hand, the unique minimal subset is $\{z\}$ and the action is not distal. Another interesting pathology is the *projection semigroup*. Let S be any space and let $M : S \times S \to S$ be projection on the first coordinate so that

$$(3.8) \qquad pq = p \qquad \text{for all} \quad p, q \in S.$$

This is a topological semigroup with every element idempotent. The translation action is minimal and equicontinuous. The action is not distal. In fact it is asymptotic. That is,

$$(3.9) \qquad PROX(S) = ASYMP(S) = S \times S.$$

It follows that the translation action is almost distal and so is semidistal.

The following relates equicontinuity and distality in general. A crucial step requires a deep theorem due to Robert Ellis.

Theorem 3.6. Let $\Phi : S \times X \to X$ be an Ellis action and let D be a dense subset of S. The following conditions are equivalent:

(1) The map $\Phi^p : X \to X$ is surjective for each $p \in D$ and $\Phi^\#(D)$ is an equicontinuous subset of X^X.
(2) The system Φ is equicontinuous and distal.
(3) The map $\Phi^p : X \to X$ is a homeomorphism for each $p \in S$ and the map $\Phi^\# : S \to \mathcal{H}(X)$ is continuous.
(4) The enveloping semigroup $\Phi^\#(S)$ is a compact topological group of homeomorphisms on X.
(5) The enveloping semigroup $\Phi^\#(S)$ is a group of homeomorphisms on X.

PROOF. (1) $\Rightarrow$ (2): By Theorem 3.5 [(3) $\Rightarrow$ (2)] the system Φ is equicontinuous and by Theorem 3.5 [(3) $\Rightarrow$ (4)] the enveloping semigroup is equal to the closure of $\Phi^\#(D)$ in $\mathcal{C}(X, X)$. By Proposition 4.2(b) $\mathcal{C}_s(X, X)$ is closed. Since $\Phi^\#(D) \subset \mathcal{C}_s(X, X)$ by assumption, it follows that $\Phi^\#(S) \subset \mathcal{C}_s(X, X)$. Now Theorem 3.1 implies that the action is distal.

(2) $\Rightarrow$ (3) and (4): By equicontinuity each Φ^p is continuous. By distality each is a bijection and so is a homeomorphism by compactness. Hence, the

image of the function $\Phi^\#$ lies in $\mathcal{H}(X)$. The function is continuous by Theorem 3.5 again. By Theorem 3.1 the enveloping semigroup is a subgroup of $\mathcal{H}(X)$ which is a topological group by Proposition 3.4(a). So the enveloping semigroup is a compact topological group of homeomorphisms.

(3) $\Rightarrow$ (1): The homeomorphisms Φ^p are surjective and equicontinuity again follows from Theorem 3.5.

(4) $\Rightarrow$ (5): This is obvious. We are simply dropping the requirement that the composition map is jointly continuous on the enveloping semigroup.

(5) $\Rightarrow$ (1): This result, which does not have an analogue in Theorem 3.3, is a special case of the celebrated Joint Continuity Theorem of Ellis. See Auslander (1988) Theorem 4.4. $\square$

Theorem 3.7. Let $\Phi : S \times X \to X$ be a point transitive Ellis action. If S is a topological semigroup then the action is topological and so is equicontinuous. If, in addition, S is a group then the action is distal and minimal.

PROOF. If x is a transitive point then $\Phi_x : S \to X$ is a surjective, continuous action map from the translation action on S to the action Φ on X. Because S is a topological semigroup the translation action is topological. It follows from Proposition 3.3(c) that Φ is a topological action and so is equicontinuous by Theorem 3.5. If S is a group then the action is also distal by Theorem 3.1. A point transitive, distal action is minimal by Corollary 2.8. $\square$

Theorem 3.8. Let $\Phi : S \times X \to X$ be an Ellis action with enveloping semigroup $\Phi^\#(S)$.
  (a) $\Phi^\#(S)$ is a topological semigroup iff for every $x \in X$ the restriction of the action to $S^+x = \{x\} \cup Sx$ is equicontinuous.
  (b) $\Phi^\#(S)$ is a topological group iff the action is distal and the restriction of the action to every minimal subset is equicontinuous.

PROOF. Notice first that the action of S is equicontinuous (or distal) iff the action of the enveloping semigroup is equicontinuous (resp. distal). Assume $\Phi^\#(S)$ is a topological semigroup. Since it acts point transitively on each S^+x it follows from Theorem 3.7 that it acts equicontinuously on each S^+x. If, in addition, $\Phi^\#(S)$ is a group then the action is distal by Theorem 3.1 and the minimal sets are exactly the sets $S^+x = Sx$. For the converse, let L_x be the enveloping semigroup of the action restricted to S^+x, i.e. $L_x =_{def} (\Phi|S^+x)^\#(S)$. $(\Phi|S^+x)^\# \circ (\Phi^\#)^{-1}$ defines a continuous, surjective homomorphism $g_x : \Phi^\#(S) \to L_x$. Furthermore, the product map

$\Pi_{x\in X} g_x : \Phi^{\#}(S) \to \Pi_{x\in X} L_x$ is injective. If the action is equicontinuous on S^+x then by Theorem 3.5 the enveloping semigroup L_x is a topological semigroup. Since $\Pi_{x\in X} g_x$ is an Ellis semigroup isomorphism of $\Phi^{\#}(S)$ onto a subsemigroup of the topological semigroup $\Pi_{x\in X} L_x$ it follows that $\Phi^{\#}(S)$ itself is a topological semigroup. If, in addition, the action is distal then by Theorem 3.1, $\Phi^{\#}(S)$ is a group and so is a topological group by Theorem 3.3(a). $\square$

REMARK 3.2. The classical example of a distal, nonequicontinuous system is given, in polar coordinates, by the map $f(r,\theta) = (r, \theta + 2\pi r)$ on the unit disc $\{r \leq 1\}$ in the plane. On each circle centered at the origin, the action is equicontinuous. For irrational radii these are minimal subsets while for rational radii these are unions of the periodic orbits which are minimal. Hence, the enveloping semigroup of the compactified action is a topological group. Since the action is not equicontinuous, the Ellis action of the enveloping semigroup is not a topological action. In fact, by Theorem 3.6 the individual bijections Φ^p are not all continuous.

The Ellis actions which are obtained by compactifying classical actions do have certain special properties. For example, such an action is *densely continuous*.

Definition 3.9. An Ellis action $\Phi : S \times X \to X$ is called *densely continuous* if there exists a dense subset D of S such that for every $p \in D$ the map $\Phi^p : X \to X$ is continuous.

If $\Phi : S \times X \to X$ is an Ellis action and A is a nonempty subset of X then $A \cup SA = S^+A$ is the smallest invariant subset of X which contains A. Its closure, however, need not be invariant. As described in Chapter 2, we define

$$(3.10) \quad \begin{aligned} [A] &=_{def} \bigcap\{K \subset X : K \text{ is closed}, A \cup SK \subset K\} \\ &= \text{the smallest closed, invariant subset of } X \text{ containing } A. \end{aligned}$$

We call $[A]$ the *closed invariant set generated by* A.

When C is a subset of the Ellis semigroup S itself then $[C]$ is the smallest closed subset of S which is invariant with respect to the translation action of S on itself. Thus, $[C]$ is the smallest closed ideal which contains C. We will then call $[C]$ the *closed ideal generated by* C.

Lemma 3.10. (a) Let $\Phi : S \times X \to X$ be an Ellis action. The set operator $[\ ]$ is a closure operator, i.e. it satisfies the following properties:
  (i) $A \subset [A]$.
  (ii) $A \subset B$ implies $[A] \subset [B]$.

(iii) $[[A]] = [A]$.
(iv) $[A \cup B] = [A] \cup [B]$.

(b) Let $\Phi : S \times X \to X$ be an Ellis action and A be a subset of X. $[A]$ contains the closure of $A \cup SA$ with equality when the action Φ is densely continuous. In particular, for a densely continuous action the closure of an invariant set is invariant.

(c) Let $\Phi : S \times X \to X$ and $\Psi : S \times Y \to Y$ be Ellis actions and $\pi : X \to Y$ be a continuous action map. For any $A \subset X$, $\pi([A]) = [\pi(A)]$.

(d) Let $\Phi : S \times X \to X$ be an Ellis action. For C any subset of S and x any point of X, $[C]x = [Cx]$.

PROOF. (a): These are easy exercises which we leave to the reader.

(b): Since $[A]$ is closed and invariant it clearly contains the closure of $A \cup SA$. To complete the proof it suffices to show that if D is a dense subset of S such that Φ^p is continuous for all $p \in D$ and $A \subset X$ is invariant then the closure $\overline{A}$ is invariant. For $p \in D$, continuity of Φ^p implies that $p\overline{A} \subset \overline{A}$. Hence, for all $x \in \overline{A}$, $Dx \subset \overline{A}$. So continuity of Φ_x and density of D imply that $S\overline{A} \subset \overline{A}$ as required.

(c): As $\pi([A])$ is a closed invariant set which contains $\pi(A)$, we have $[\pi(A)] \subset \pi([A])$. On the other hand, if L is any closed invariant subset of Y which contains $\pi(A)$, then A is a subset of the closed invariant set $\pi^{-1}(L)$. Hence, $[A] \subset \pi^{-1}(L)$ and so $\pi([A]) \subset L$. Intersecting over all such L we have $\pi([A]) \subset [\pi(A)]$.

(d): Apply (c) to the action map $\Phi_x : S \to X$. □

As we will not be assuming that our actions are densely continuous, $[A]$ might be strictly larger than the closure of $A \cup SA$. For example, the min-center $[Min(X)]$ might be strictly larger than the closure of the invariant set $Min(X)$.

We conclude this section with a useful, related quotient space result.

Proposition 3.11. Let $\Phi : S \times X \to X$ be an Ellis action and D be a dense subset of S. Assume that $\pi : X \to Y$ is a surjective continuous map of spaces such that for every $p \in D$:

(3.11) $\qquad \pi(x_1) = \pi(x_2) \quad \Longrightarrow \quad \pi(px_1) = \pi(px_2)$.

There is then a unique semigroup action $\Psi : S \times Y \to Y$ such that π is an action map. Furthermore, Ψ is an Ellis action.

PROOF. Because D is dense in S and $p \mapsto px$ is continuous for $x \in X$, it follows that (3.11) holds for all $p \in S$. It follows that there is a unique, well-defined function $\Psi : S \times Y \to Y$ such that for $y = \pi(x)$

(3.12) $$\Psi_y \;=\; \pi \circ \Phi_x.$$

or, equivalently, for all $p \in S$ we have $py = \pi(px)$.

Hence, for $p, q \in S$

(3.13) $$p(qy) \;=\; p(\pi(qx)) \;=\; \pi(p(qx)) \;=\; \pi((pq)x) \;=\; (pq)y$$

and so Ψ is a semigroup action. Equation (3.12) implies that it is an Ellis action, i.e. that each Ψ_y is continuous. $\square$

CHAPTER 4

Applications Using Ideals

Let $\Phi : S \times X \to X$ be an Ellis action. If H is a closed subsemigroup of the Ellis semigroup S then the restriction of Φ to $H \times X$ is an Ellis action of H on X. Furthermore, an action map of S actions is an action map of the induced H actions.

This restricted action is the pullback via the inclusion map $g : H \to S$. See Lemma 1.5. As in the proof of Lemma 1.5 we will write $TRANS_H$, $RECUR_H$, $PROX_H$, etc. for those points in X or pairs in $X \times X$ which are transitive, recurrent or proximal, etc. with respect to the action of H.

If Φ is an S action or π an action map of S actions then we will say that Φ or π is H asymptotic, H proximal, H distal, H almost distal or H semidistal if it satisfies the corresponding property regarded as an H action or as an action map of H actions. Using the Ellis action approach we discover several interesting properties which are related by moving about among subsemigroups. For the various properties of S actions we will drop the label only for S itself. For example, a subset is called invariant when it is S invariant, not merely H invariant.

We first observe that for distality, the concept remains unchanged when we restrict to a closed ideal.

Lemma 4.1. Let $\Phi : S \times X \to X$ be an Ellis action and let L be a closed ideal in S.

(a) If $x \in X$, then

$$(4.1) \qquad x \in Lx \implies Sx = Lx.$$

(b) A point is transitive for the restriction $\Phi|L$ iff it is a transitive point which is L recurrent. That is,

$$(4.2) \qquad TRANS_L = TRANS_S \cap RECUR_L$$

(c) A point is minimal for the restriction $\Phi|L$ iff it is a minimal point for Φ.

(d) The restriction $\Phi|L$ is minimal iff Φ is minimal.

PROOF. (a): Because L is an ideal, i.e. $SL \subset L$, it follows that Lx is invariant. Hence, $x \in Lx$ implies $Sx \subset Lx$. Since $L \subset S$, the reverse inclusion is obvious.

(b): If $x \in RECUR_L$ then $x \in Lx$ and so (4.1) implies that $Lx = Sx$ which equals X if x is also a transitive point. Hence, $TRANS_S \cap RECUR_L \subset TRANS_L$. The reverse inclusion is obvious.

(c): L contains some minimal ideal J. A point x is minimal iff $Jx = Sx$ and $x \in Sx$. A point is L minimal iff $Jx = Lx$ and $x \in Lx$. The first implies the second since $J \subset L \subset S$. The converse follows from (a).

(d): The action is minimal iff every point is minimal and transitive. Apply (b) and (c). □

Proposition 4.2. Let $\Phi : S \times X \to X$ and $\Psi : S \times Y \to Y$ be Ellis actions and $\pi : X \to Y$ be a continuous action map with Y minimal. Let L be a closed ideal of S.
  (a) Φ is distal iff it is L distal. In that case, $\Phi^\#(S) = \Phi^\#(L)$, i.e. the enveloping semigroups agree..
  (b) π is distal iff it is L distal.

PROOF. (a): By Lemma 4.1(c) every point of $X \times X$ is L minimal iff every point is minimal. By Theorem 2.7(a) Φ is L distal iff it is distal. In any case, $\Phi^\#(L) \subset \Phi^\#(S)$. By Theorem 2.7(a) again, L distality implies that $\Phi^\#(L)$ is a group of bijections with identity $1_X = \Phi^u$ for any idempotent $u \in L$. For any $p \in S, pu \in L$ and so $\Phi^p = \Phi^p \circ \Phi^u = \Phi^{pu} \in \Phi^\#(L)$. Hence, $\Phi^\#(S) \subset \Phi^\#(L)$.

(b): Use Corollary 2.11, applying (a) to the restriction of the product action to R_π. □

REMARK 4.1. For any closed subsemigroup H, the inclusion $PROX_H \subset PROX$ shows that distality for Φ or π implies H distality even without the assumption that Y is minimal.

We will now consider the meaning of these H notions when H is $[Min(S)]$, $[Id(S)]$ or $[Id(S)S]$ the closed ideals generated by the minimal points, by the idempotents or by the recurrent points of S (see equation (2.16)). Notice that every minimal point of S is in an ideal generated by a minimal idempotent and so

(4.3) $\quad [Id(Min(S))] \;=\; [Min(S)] \;\subset\; [Id(S)] \;\subset\; [Id(S)S]$.

Furthermore, every ideal contains minimal idempotents and so meets $[Min(S)]$.

For an Ellis action $\Phi : S \times X \to X$ the concepts of minimal set and minimal point are the same for the S action as for the restricted $[Id(S)S]$, $[Id(S)]$ and $[Min(S)]$ actions. This follows from Proposition 2.2(b) and the fact the each of these sets contain all the minimal idempotents. Similarly, the concepts of recurrent point and transitive point are the same for the S action as for the restricted $[Id(S)S]$ and $[Id(S)]$ actions by Proposition 2.4(a) and Lemma 4.1(b). However, the $[Min(S)]$ versions of the latter concepts are sharper and are worth remarking.

Proposition 4.3. For an Ellis action $\Phi : S \times X \to X$, let K be a closed invariant subset of X and x be a point of X.

(a) $\quad Id(S)K \;=\; Id(S)SK$
$\;=\; RECUR(K) \;=\; K \cap RECUR(X).$

(b) $\quad Id(Min(S))K \;=\; Min(K) \;=\; K \cap Min(X).$

(c) The $[Id(S)S]$ orbit of x, $[Id(S)S]x$ is the smallest closed invariant subset of the orbit Sx which contains all the recurrent points in the orbit. That is,

(4.4)
$$[Id(S)S]x \;=\; [Id(S)Sx] \;=\; [RECUR(Sx)]$$
$$=\; [Sx \cap RECUR(X)].$$

(d) The $[Min(S)]$ orbit of x, $[Min(S)]x$ is the smallest closed invariant subset of the orbit Sx which contains all the minimal points in the orbit. That is,

(4.5)
$$[Min(S)]x \;=\; [Min(S)x] \;=\; [Min(Sx)]$$
$$=\; [Sx \cap Min(X)].$$

PROOF. (a): The first equation is just equation (2.16) applied to K. On the one hand for any idempotent u, $uux = ux$ implies that ux is recurrent. On the other hand, Proposition 2.4(a) implies that for any recurrent point x, $ux = x$ for some idempotent u. A recurrent point of K is just a point of K which is fixed by some idempotent, i.e. a point of K which is recurrent in X.

(b): This repeats the argument of (a) replacing "idempotent" by "minimal idempotent" and Proposition 2.4(a) by Proposition 2.4(b).

(c): For the first equation of (4.4) apply Lemma 3.10(d) with $C = Id(S)S$ of the domain. Notice that for any nonempty $C \subset S$, $[C]$ a closed ideal implies that $[C]x = [Cx]$ is a closed invariant set. The other two equations follow from part (a) with $K = Sx$.

(d): For the first equation of (4.5) apply Lemma 3.10(d) again. For the second equation, apply equation (2.1) to Φ_x to get

(4.6) $\qquad Min(S)x \;=\; Min(Sx).$

Finally, the third equation follows from part (b) with $K = Sx$. $\square$

REMARK 4.2. It follows from part (a) that $Id(S)S$ is the subset $RECUR(S)$, i.e. the set of points which are recurrent with respect to the translation action of S on itself.

Recall that $[Id(S)X] = [RECUR]$ is called the *center* of X and $[Min(X)]$ is called the *min-center* of X. So (4.4) and (4.5) say that $[Id(S)S]x$ and $[Min(S)]x$ are, respectively, the center and the min-center of the orbit subsystem Sx. Notice especially that these sets might be strictly smaller than $Sx \cap [RECUR]$ and $Sx \cap [Min(X)]$. A point could be a limit of recurrent points which are not in the orbit of x without being near any recurrent points of the orbit.

On the other hand, $Id(S)x$ consists of recurrent points which are proximal to x and so is usually a proper subset of the set $Id(S)Sx$ of recurrent points in the orbit of x.

We say that a point $x \in X$ is $[Min(S)]$ *recurrent* when it is recurrent for the restricted action, i.e. $x \in [Min(S)]x$. $RECUR_{[Min(S)]}$ is the set of $[Min(S)]$ recurrent points. From Lemma 4.1 and equation (4.5) we have:

(4.7)
$$x \in [Min(S)]x \implies Sx = [Min(S)]x = [Min(Sx)].$$
$$TRANS_{[Min(S)]} = TRANS_S \cap RECUR_{[Min(S)]}.$$

As with any closed subsemigroup, we call the system $[Min(S)]$ *point transitive* when $TRANS_{[Min(S)]} \neq \emptyset$.

Proposition 4.4. Let $\Phi : S \times X \to X$ be an Ellis action.

(a) A point $x \in X$ is $[Min(S)]$ recurrent iff it is S recurrent and the min-center of the orbit Sx is the entire orbit.

(b) The action Φ is $[Min(S)]$ point transitive iff $TRANS \neq \emptyset$ and $X = [Min(X)]$, i.e. it is point transitive as an S action and the min-center is the whole space.

(c) A pair $z \in X \times X$ is $[Min(S)]$ asymptotic iff its entire orbit consists of proximal pairs, i.e. $Sz \subset PROX$.

PROOF. (a): This says, in our notation:

(4.8) $\quad x \in [Min(S)]x \iff x \in Sx$ and $Sx = [Min(Sx)]$.

If $x \in Sx = [Min(Sx)]$ then equation (4.5), $[Min(Sx)] = [Min(S)]x$, implies $x \in [Min(S)]x$. The reverse implication follows from (4.7).

(b): If the action is point transitive and the min-center is the whole space then for $x \in TRANS$, $Sx = X = [Min(X)] = [Min(Sx)]$. By (4.5) the latter set is $[Min(S)]x$ and so x is a $[Min(S)]$ transitive point.

On the other hand, if x is a $[Min(S)]$ transitive point, i.e. $X = [Min(S)]x$, then x is surely a transitive point and a $[Min(S)]$ recurrent point. By part (a) the mincenter of the orbit is the entire orbit. Since x is a transitive point the orbit is the entire space. Thus, the min-center is the whole space.

(c): The pair z is $[Min(S)]$ asymptotic iff $Min(S) \subset Foc_z$, i.e. iff $pz \in 1_X$ for every minimal $p \in S$. This implies $pqz \in 1_X$ for any $q \in S$ since pq is minimal as well. That is, $qz \in PROX$ for all q.

4. APPLICATIONS USING IDEALS

On the other hand, if $pz \not\in 1_X$ for some minimal p, then the minimal invariant set Spz is disjoint from 1_X and so pz is an element of the orbit of z which is not proximal. $\square$

For an Ellis action $\Phi : S \times X \to X$ we define the set

(4.9)
$$LPROX(\Phi) =_{def} \{(x_1, x_2) \in X \times X : (px_1, px_2) \in PROX \text{ for all } p \in S\}$$
$$\subset PROX(\Phi).$$

For classical $\mathbb{Z}_+$ actions on a metric space this is called the *syndetic proximality relation* because a pair z lies in $LPROX$ for such an action when for every positive ϵ the set of times spent ϵ close to the diagonal is a syndetic subset of $\mathbb{Z}_+$ (See Glasner and Maon (1989)). The inclusion $LPROX \subset PROX$ holds because $pz \in PROX$ for any $p \in S$ implies $z \in PROX$ (this is the *capturing property* described below).

We see from Proposition 4.4 (c) that

(4.10) $\qquad LPROX(\Phi) \quad = \quad ASYMP_{[Min(S)]}(\Phi).$

Furthermore, by (2.9) and (4.6) $[Min(S)]p = [Min(S)p] \subset [Min(S)]$ for all $p \in S$ and so

(4.11) $\qquad S[Min(S)]S \quad \subset [Min(S)].$

That is, $[Min(S)]$ is a closed, two-sided ideal. It then follows from Proposition 1.10 that $L = ASYMP_{[Min(S)]}$ is an invariant equivalence relation.

The relation $LPROX$ was introduced by Clay in Clay (1963). He there proved $LPROX$ is an invariant equivalence relation, i.e. the result we have just recovered.

A convenient way of comparing $PROX$ and $LPROX$ uses the equivalence relations $(\Phi^u)^{-1}(1_X) = \{(x,y) : ux = uy\}$ for all minimal idempotents u. From Proposition 2.4(c) we see that:

(4.12) $\qquad PROX \quad = \quad \bigcup \{(\Phi^u)^{-1}(1_X) : u \in Id(Min(S))\}.$

On the other hand, for any pair $z \in X \times X$, Foc_z is a closed ideal. Hence, if it contains every minimal idempotent then it contains all of $[Min(S)]$. It follows that

(4.13) $\qquad LPROX \quad = \quad \bigcap \{(\Phi^u)^{-1}(1_X) : u \in Id(Min(S))\}.$

We now describe the distality concepts associated with the actions restricted to these ideals.

Theorem 4.5. Let $\Phi : S \times X \to X$ and $\Psi : S \times Y \to Y$ be Ellis actions and $\pi : X \to Y$ be a continuous action map.

(a) For the action map π distality, $[Id(S)S]$ distality, $[Id(S)]$ distality and $[Min(S)]$ distality are equivalent conditions.

(b) For the action map π semidistality, $[Id(S)S]$ semidistality and $[Id(S)]$ semidistality are equivalent conditions.
(c) The following are equivalent.
  (i) The map π is proximal.
  (ii) The map π is $[Id(S)S]$ proximal.
  (iii) The map π is $[Id(S)]$ proximal.
  (iv) The map π is $[Min(S)]$ proximal.
  (v) The map π is $[Min(S)]$ asymptotic.
  (vi) Every minimal subset of R_π is contained in 1_X, i.e. $Min(R_\pi) \subset 1_X$.
  (vii) $PROX(\Phi) = (\pi \times \pi)^{-1}(PROX(\Psi))$.
  (viii) $R_\pi \subset LPROX(\Phi)$.
  When these conditions hold $\Phi^p : X \to X$ is constant on the fiber $\pi^{-1}(y)$ for every $y \in Y$ and every $p \in [Min(S)]$.
(d) The following are equivalent.
  (i) The map π is $[Id(S)S]$ asymptotic.
  (ii) The map π is $[Id(S)]$ asymptotic.
  (iii) Every recurrent point of R_π is contained in 1_X, i.e. $RECUR(R_\pi) \subset 1_X$.

PROOF. (a): This is a special case of Proposition 4.2.
(b): By Proposition 2.4(a) a point is recurrent iff it is $[Id(S)]$ recurrent, i.e.
$$(4.14) \qquad RECUR = RECUR_{[Id(S)S]} = RECUR_{[Id(S)]}.$$
Similarly, by Proposition 2.4(c)
$$(4.15) \qquad \begin{aligned} PROX &= PROX_{[Id(S)S]} \\ &= PROX_{[Id(S)]} = PROX_{[Min(S)]}. \end{aligned}$$
Together these yield (b).

(c),(i) $\Leftrightarrow$ (ii) $\Leftrightarrow$ (iii) $\Leftrightarrow$ (iv): Apply Lemma 1.5(c), or just use (4.13).

(i)$\Rightarrow$(vi): If $M \subset R_\pi$ is minimal and is not contained in 1_X then it is disjoint from 1_X. If $z \in M$ then $pz \in M$ for all $p \in S$ and so $z \notin PROX$. Contrapositively, $R_\pi \subset PROX$ implies (vi).

(vi)$\Rightarrow$(v): If $p \in Min(S)$ then Sp is a minimal ideal with $p \in S$. If $z \in R_\pi$ then Spz is a minimal subset of R_π and so assumption (vi) implies $Spz \subset 1_X$. Hence, $MIN(S)$ is contained in the closed ideal Foc_z which therefore contains $[Min(S)]$ as well.

(v)$\Rightarrow$(iv): Obvious.
(i) $\Leftrightarrow$ (vii): This is Lemma 1.6(c).
(v) $\Leftrightarrow$ (viii): This is clear from (4.10).

Assuming these conditions, R_π is contained in $ASYMP_{[Min(S)]}$ by (v). It then follows from Proposition 1.10 that for every $p \in [Min(S)]$ the map Φ^p is constant on each fiber of π.

4. APPLICATIONS USING IDEALS

(d)(i)⇒(ii): Obvious.

(ii)⇒(iii): If $z \in R_\pi$ is recurrent then assumption (ii) implies that $z \in ASYMP \cap RECUR^2$ for the $[Id(S)]$ action. So by (1.13) $z \in 1_X$.

(iii)⇒(i): If $z \in R_\pi$, u is an idempotent and $p \in S$ then upz is a recurrent point of R_π and so by assumption (iii) $upz \in 1_X$. Since Foc_z is a closed ideal, $[Id(S)S] \subset Foc_z$. □

REMARK 4.3. Theorem 2.9(e) can now be interpreted as saying that if π is semidistal and proximal, i.e. $[Min(S)]$ asymptotic, then it is $[Id(S)S]$ asymptotic.

Theorem 4.6. Let $\Phi : S \times X \to X$ be an Ellis action.
  (a) The following properties are equivalent
    (i) Φ is $[Min(S)]$ almost distal.
    (ii) $PROX(\Phi) = ASYMP_{[Min(S)]}(\Phi)$.
    (iii) $PROX(\Phi) = LPROX(\Phi)$.
    (iv) For each point z in $X \times X$ the orbit Sz with respect to the product system Φ^2 contains a unique minimal set.
    (v) The enveloping semigroup $\Phi^\#(S)$ contains a unique minimal ideal.
    (vi) $PROX(\Phi)$ is an equivalence relation on X.
    (vii) $PROX(\Phi)$ is an invariant subset of $X \times X$.
  If Φ is $[Min(S)]$ almost distal then for each point $x \in X$ the orbit Sx contains a unique minimal set.
  (b) The following properties are equivalent
    (i) Φ is $[Min(S)]$ semidistal.
    (ii) Each $[Min(S)]$ recurrent point in $X \times X$ is minimal.
  If Φ is $[Min(S)]$ semidistal then each $[Min(S)]$ recurrent point in X is minimal.
  (c) The following properties are equivalent
    (i) Φ is $[Id(S)S]$ almost distal.
    (ii) Φ is $[Id(S)]$ almost distal.
    (iii) For each point z in $X \times X$ the center of the orbit Sz is a minimal invariant set.
    (iv) Φ is semidistal and $[Min(S)]$ almost distal.
    (v) Φ is semidistal and $PROX(\Phi)$ is an equivalence relation on X.
  If Φ is $[Id(S)]$ almost distal then for each point x in X the center of the orbit Sx is a minimal invariant set.

PROOF. (a): Notice first that if the mincenter $[Min(X)]$ is a minimal invariant set then it is the unique minimal invariant subset of X. Conversely, if K is the unique minimal subset of X then $[Min(X)] = K$. Also, (2.2) implies that $Min(\Phi^\#(S)) = \Phi^\#(Min(S))$. Hence, the equivalence of (i),(iv) and (v) as well as the concluding remark are restatements of Theorem 2.7(b) applied to the $[Min(S)]$ action.

(i) $\Leftrightarrow$ (ii): Since $PROX = PROX_{[Min(S)]}$, this is the definition of almost distality for the $[Min(S)]$ action.

(ii) $\Leftrightarrow$ (iii): This follows from (4.10).

(ii) $\Rightarrow$ (vii): $[Min(S)]$ is a closed, two-sided ideal by (4.11). So $PROX = ASYMP_{[Min(S)]}$ is invariant by Proposition 1.10.

(vii) $\Rightarrow$ (vi): Condition (vii) says that $PROX$ equals the subset L defined by (4.9). Then (4.10) and Proposition 1.10 imply that $PROX$ is an equivalence relation.

It is a bit more enlightening to prove this directly: If $(x_1, x_2), (x_2, x_3) \in PROX$ then there exists $p \in S$ such that $px_1 = px_2$. From condition (vii), $(px_2, px_3) \in PROX$ and so there exists $q \in S$ such that $qpx_3 = qpx_2 = qpx_1$ and so $(x_1, x_3) \in PROX$.

(vi) $\Rightarrow$ (v): Assume that J, K are minimal ideals in S and that $u \in Id(J)$. By minimality $Ku = J$. With respect to the translation action $Iso_u \cap K$ is a nonempty closed subsemigroup and so it contains an idempotent $v \in K$ which satisfies $vu = u$. For any $x \in X$, the pairs $(x, ux), (x, vx)$ are in $PROX$. By assumption (vi) $(ux, vx) \in PROX$. There exists $p \in S$ such that $pux = pvx$. Since K is a minimal ideal, there exists $q \in S$ such that $qpv = v$, and so $ux = vux = qpvux = qpux = qpvx = vx$. Since $ux = vx$ for all $x \in X$, $\Phi^\#(v) = \Phi^\#(u)$ and so the ideals J and K have the same image under $\Phi^\#$. Thus, the image of $Min(S)$ is the image of any minimal ideal J in S.

(b): This is a restatement of Theorem 2.7(c) applied to the $[Min(S)]$ action.

(c) (i) $\Rightarrow$ (ii): By (4.10) the two notions of proximality agree. Asymptotic with respect to $[Id(S)S]$ implies asymptotic with respect to the smaller ideal $[Id(S)]$.

(i) $\Leftrightarrow$ (iii): By Theorem 2.7(b) applied to the $[Id(S)S]$ action, together with Proposition 4.3(c), (i) is equivalent to (iii) and implies the remark at the end.

(ii) $\Rightarrow$ (iv) $\Rightarrow$ (i): By Theorem 2.7(b) applied to the $[Id(S)]$ action, (i) is equivalent to the condition that $\Phi^\#([Id(S)])$ is minimal. If $\Phi^\#([Id(S)])$ minimal then the enveloping semigroup $\Phi^\#(S)$ contains a unique minimal ideal and that this ideal contains all the idempotents. That is (1) $\Phi^\#(S)$ contains a unique minimal ideal, and (2) Every idempotent of $\Phi^\#(S)$ is minimal. Condition (1) is equivalent to $[Min(S)]$ almost distality by part (a) and condition (2) is equivalent to semidistality by Theorem 2.7(c). Thus, (ii) implies (iv). On the other hand, if every idempotent of $\Phi^\#(S)$ is minimal, i.e. $Id(\Phi^\#(S)) \subset Min(\Phi^\#(S))$ then by (2.9) applied to the enveloping semigroup $Id(\Phi^\#(S))\Phi^\#(S) \subset Min(\Phi^\#(S))\Phi^\#(S) \subset Min(\Phi^\#(S))$ as well.

By Proposition 4.3(a) every recurrent point in $\Phi^{\#}(S)$ is minimal and so it follows that $\Phi^{\#}([Id(S)S])$ is contained in this unique minimal ideal. Thus, (iv) implies (i) by Theorem 2.7(b) again applied to the $[Id(S)S]$ action. (iv) $\Leftrightarrow$(v). By part (a). $\square$

REMARK 4.4. It follows from Proposition 1.10 that when $PROX$ is an equivalence relation then for every $p \in [Min(S)]$ the map $\Phi^p : X \to X$ is constant on each equivalence class.

For the case of $\mathbb{Z}_+$ actions we have organized the distality properties associated with these different ideals into a table which has been placed at the end of this book as a reward for readers who get that far.

As part (a) of Theorem 4.6 indicates, $PROX$ is not always an invariant subset. It does, however, always satisfy a useful negative invariance condition. We say that a subset A of X satisfies the *capturing property* if $px \in A$ for some $p \in S$ implies $x \in A$ (see Auslander and Glasner (2002)). Notice that this is equivalent to the condition that the complement $X \setminus A$ is invariant. For example, if $Spx = X$ then $Sx = X$, i.e. TRANS satisfies the capturing property. Similarly, if Spz meets 1_X for $z \in X \times X$ then Sz meets 1_X. That is, $PROX$ satisfies the capturing property in $X \times X$.

An asymptotic map is clearly almost distal. Hence, from Theorem 4.5(c) we see that for any S action or action map of S actions
(4.16)
$$\text{proximal} \implies [Min(S)] \text{ almost distal} \implies [Min(S)] \text{ semidistal}$$

Theorem 4.7. Let $\Phi : S \times X \to X$ and $\Psi : S \times Y \to Y$ be Ellis actions and $\pi : X \to Y$ be a continuous action map. Assume that Y is minimal. The action map π is $[MIN(S)]$ semidistal iff each $[Min(S)]$ recurrent point in R_π is minimal.

PROOF. Apply Corollary 2.11(b) to the $[Min(S)]$ action. $\square$

While for almost distality in general this kind of reduction from map to space results can't be done, it can for $[Min(S)]$ almost distality.

Theorem 4.8. Let $\Phi : S \times X \to X$ and $\Psi : S \times Y \to Y$ be Ellis actions and $\pi : X \to Y$ be a continuous action map.
  (a) If the action map π is proximal and the action Ψ is $[Min(S)]$ almost distal then the action Φ is $[Min(S)]$ is. On the other hand, if the action map π is surjective and the action Φ is $[Min(S)]$ almost distal then the action Ψ is $[Min(S)]$ almost distal as well.
  (b) Assume that Y is minimal. The following are equivalent.

(i) The action map π is $[Min(S)]$ almost distal.
(ii) $PROX(\Phi) \cap R_\pi = ASYMP_{[Min(S)]}(\Phi) \cap R_\pi$.
(iii) For every $y \in Y$ the $Iso_y \cap [Min(S)]$ action on the fiber X_y, $\Phi(y) : Iso_y \times X_y \to X_y$ is almost distal.
(iv) For every $y \in Y$ the Iso_y action on the fiber X_y, $\Phi(y) : Iso_y \times X_y \to X_y$ is $[Min(Iso_y)]$ almost distal.
(v) For each point z in R_π the orbit Sz with respect to the product system Φ^2 contains a unique minimal set.
(vi) $PROX(\Phi) \cap R_\pi$ is an equivalence relation on X.
(v) $PROX(\Phi) \cap R_\pi$ is an invariant subset of $X \times X$.

(c) Assume that $\Theta : S \times Z \to Z$ is a minimal Ellis action and that $\epsilon : Y \to Z$ is a continuous action map. If the action map π is proximal and the action map ϵ is $[Min(S)]$ almost distal then the composition $\epsilon \circ \pi$ is a $[Min(S)]$ almost distal action map. On the other hand, if the action map π is surjective and the composition $\epsilon \circ \pi$ is $[Min(S)]$ almost distal then the action map ϵ is a $[Min(S)]$ almost distal action map.

PROOF. (a): Since π is proximal, Theorem 4.5(c) implies $PROX(\Phi) = (\pi \times \pi)^{-1}(PROX(\Psi))$. So $PROX(\Phi)$ is an equivalence relation if $PROX(\Psi)$ is. Apply Theorem 4.6(a). The converse result follows from Theorem 2.9(b).

(b): Notice that π is surjective and every point of Y is recurrent because Y is minimal. As in the proof of Corollary 2.11, if $y \in Y$ then Iso_y is a closed co-ideal which meets every ideal of S and so Lemma 2.2(e)(iii) holds for $H = Iso_y$.

(i) $\Leftrightarrow$ (ii): Obvious.

(i) $\Rightarrow$ (iii): Proposition 1.4.

(iii) $\Rightarrow$ (iv): While Lemma 2.2(e) says that $Iso_y \cap Min(S) = Min(Iso_y)$ we only get that $Iso_y \cap [Min(S)]$ contains $[Min(Iso_y)]$ where the latter is the smallest closed Iso_y ideal which contains $Min(Iso_y)$. The inclusion says that asymptotic with respect to $Iso_y \cap [Min(S)]$ implies asymptotic for $[Min(Iso_y)]$. Since the notion of proximality on X_y is the same for both, it follows that (iii) implies (iv).

(iv) $\Rightarrow$ (i): Assume that $(x_1, x_2) \in PROX(\Phi) \cap R_\pi$ with $y = \pi(x_1) = \pi(x_2)$. Let J be any minimal ideal in S. By Lemma 2.2(e) $J \cap Iso_y$ is a minimal Iso_y ideal. By assumption (iii), $J \cap Iso_y \subset Foc_{(x_1,x_2)}$. Since $J = S(J \cap Iso_y)$ by minimality of J and since $Foc_{(x_1,x_2)}$ is an S ideal we have $J \subset Foc_{(x_1,x_2)}$. As this is true for every minimal ideal J and $Foc_{(x_1,x_2)}$ is closed, $[Min(S)] \subset Foc_{(x_1,x_2)}$ as required.

In order to prove the remaining equivalences we observe that Proposition 2.4(c) implies

(4.17) $\qquad PROX(\Phi_y) \quad = \quad PROX(\Phi) \cap X_y \times X_y.$

Furthermore,

(4.18)
$$p \in S \text{ and } z, pz \in X_y \times X_y \implies p \in Iso_y,$$
$$\text{So} \quad z \in X_y \times X_y \implies Iso_y z = Sz \cap (X_y \times X_y).$$

Now we apply the equivalences Theorem 4.6(a) to the Iso_y action on X_y.

(iv) $\Leftrightarrow$ (v): By Theorem 4.6(a), (iv) holds iff for all $y \in Y$ and $z \in X_y \times X_y$ the orbit $Iso_y z$ contains a unique Iso_y minimal subset. We show that this is equivalent to the condition that Sz contains a unique S minimal subset. Assume N is the unique Iso_y minimal subset. Let M be any minimal subset of Sz. Since Y is minimal, M meets $X_y \times X_y$. Minimality and uniqueness of N imply $N \subset M$. By minimality of M, $SN = M$ and so M is uniquely determined by N. On the other hand, assume M is the unique minimal subset of Sz and that N is any Iso_y minimal subset of $X_y \times X_y$. The ideal SN in Sz must contain M by uniqueness. By (4.16) $SN \cap (X_y \times X_y) = Iso_y N = N$. By minimality of N, $M \cap (X_y \times X_y) = N$ and so N is uniquely determined by M.

(iv) $\Leftrightarrow$ (vi): Since R_π is an equivalence relation with equivalence classes the X_y's, (4.15) implies that $PROX \cap R_\pi$ is an equivalence relation on X iff for every $y \in Y$, $PROX \cap (X_y \times X_y)$ is an equivalence relation on X_y. So the result follows from Theorem 4.6(a) again.

(iv) $\Leftrightarrow$ (vii). By (4.15) and Theorem 4.6(a) again (iii) holds iff each $PROX \cap (X_y \times X_y)$ is Iso_y invariant. Clearly, it is sufficient that $PROX \cap R_\pi$ be S invariant. On the other hand, if $z \in PROX \cap (X_y \times X_y)$ and $p \in S$ then there exists $q \in S$ such that $qpz \in X_y \times X_y$. Since $qp \in Iso_y$, Iso_y invariance will imply that $qpz \in PROX$. It follows that pz is in $PROX$ because $PROX$ has the capturing property.

(c): Mimic the proof of Proposition 2.12(a) using part (b) to derive the map results of (c) from the space results of (a). $\square$

Theorem 4.9. Let $\Phi : S \times X \to X$ and $\Psi : S \times Y \to Y$ be Ellis actions and $\pi : X \to Y$ be a surjective continuous action map. Assume that X is $[Min(S)]$ point transitive. If π is proximal, or, more generally, if π is $[Min(S)]$ semidistal, then Y is point transitive and π is minimal. If, in addition, Y is minimal then X is minimal.

PROOF. This is just Theorem 2.14 applied to the induced $[Min(S)]$ action. As noted in (4.14), if π is proximal then it is $[Min(S)]$ semidistal. $\square$

REMARK 4.5. Suppose that $\Phi : S \times X \to X$ is an Ellis action such that the point $m \in X$ is a fixed point, i.e. $Sm = m$, and the singleton m is the unique minimal subset of X. Then (m, m) is the unique minimal subset of

$X \times X$ and so $PROX = X \times X$, i.e. the action Φ is proximal. The system is called *point weak mixing* if Φ^2 is point transitive. If Φ is nontrivial, point weak mixing and has a unique fixed point minimal subset, then the first coordinate projection $\pi : X \times X \to X$ is a surjective, proximal action map on a point transitive system which is nonetheless not a minimal map since the diagonal 1_X is a proper invariant subset which maps onto X. Thus, the condition of $[Min(S)]$ point transitivity is necessary for the result. There exist classical dynamical systems which are nontrivial, point weak mixing and with a unique fixed point minimal subset. See, for example, the stopped torus example in Akin (1993) Chapter 9.

A continuous surjective map $\pi : X \to Y$ of spaces is called *irreducible* if the only closed subset K of X such that $\pi(K) = Y$ is X itself. For any Ellis semigroup S and any space X, the second coordinate projection map $\pi_2 : S \times X \to X$ is an Ellis action which we will call the *identity action* of S on X. Since $px = x$ for all $p \in S$, the enveloping semigroup is just $\{1_X\}$. Any continuous map between spaces is an action map of the corresponding identity actions. Clearly a space map is irreducible iff it is a minimal map of the identity actions.

Define

(4.19) $\quad Inj(\pi) \ =_{def} \ \{x \in X : \pi^{-1}(\pi(x)) \text{ is a singleton}\}.$

When $Inj(\pi)$ is dense in X, the map π is called an *almost one-to-one* map.

Proposition 4.10. Let $\pi : X \to Y$ and $\epsilon : Y \to Z$ be continuous surjective maps of spaces.

(a) The map π is irreducible iff for every nonempty open $U \subset X$ there exists a nonempty open $V \subset Y$ such that $\pi^{-1}(V) \subset U$.
(b) If π is almost one-to-one, i.e. the set $Inj(\pi)$ is dense in X, then π is irreducible. If π is irreducible and X is a metric space then $Inj(\pi)$ is a dense G_δ subset of X and, in addition, $\pi(Inj(\pi))$ is a dense G_δ subset of Y. Thus, when X is metrizable, π is irreducible iff it is almost one-to-one.
(c) Assume that π is irreducible. If $K \subset R_\pi$ is closed and $(\pi \times \pi)(K) = 1_Y$ then $K \supset 1_X$.
(d) The composed map $\epsilon \circ \pi$ is irreducible iff both ϵ and π are irreducible maps.

PROOF. (a): If π is irreducible and $U \subset X$ is nonempty and open then $K = X \setminus U$ is a proper closed subset of X. Hence, $\pi(K)$ is a proper closed subset of Y. $V = Y \setminus \pi(K)$ is a nonempty open set and $\pi^{-1}(V) \subset U$. On the other hand, if $K \subset X$ and $\pi(K) = Y$ then K meets $\pi^{-1}(V)$ for every nonempty subset V of Y. So the open set condition implies that K is dense. If K is closed then $K = X$.

(b): If $\pi(K) = Y$ then $K \supset Inj(\pi)$. If K is closed and Inj is dense then $K = X$. Assume that π is irreducible and X admits a metric d. For every $x \in X$ and $\epsilon > 0$ there is a nonempty open $V \subset Y$ such that $\pi^{-1}(V) \subset V_{\epsilon/2}(x)$ and so $\pi^{-1}(V) \times \pi^{-1}(V) \subset V_\epsilon = \{(x_1, x_2) : d(x_1, x_2) < \epsilon\}$. Hence, $\{x \in X : \pi^{-1}(\pi(x)) \times \pi^{-1}(\pi(x)) \subset V_\epsilon\}$ is open and dense in X. Intersect over positive rational ϵ to get Inj and apply the Baire Category Theorem. Similarly, intersecting the open sets $\{y \in Y : \pi^{-1}(y) \times \pi^{-1}(y) \subset V_\epsilon\}$ shows that the dense image of Inj is a G_δ as well.

(c): Assume π is irreducible, $K \subset R_\pi$ is closed and $(\pi \times \pi)(K) = 1_Y$. For nonempty open $U \subset X$ there exists a nonempty open $V \subset Y$ such that $\pi^{-1}(V) \subset U$. If $(x_1, x_2) \in K$ with $\pi(x_1) = \pi(x_2) \in V$ then $x_1, x_2 \in U$. Hence, $U \times U$ meets K. Since K is closed $K \supset 1_X$.

(d): Use the proof of Proposition 2.13(a) or just apply the result itself to the identity actions. $\square$

REMARK 4.6. Kolyada et al (2001) show that if a continuous map $f : X \to X$ defines a minimal action of $\beta\mathbb{Z}_+$ then f is an irreducible map.

Proposition 4.11. Let $\Phi : S \times X \to X$ and $\Psi : S \times Y \to Y$ be Ellis actions and $\pi : X \to Y$ be a surjective continuous action map. If π is an irreducible map then it is a minimal action map. If, in addition, Y is minimal then π is proximal.

PROOF. It is obvious that an irreducible map π is minimal. Now assume that Y is minimal and that $M \subset R_\pi$ is minimal. $(\pi \times \pi)(M)$ is an invariant subset of, and so equals, the minimal set 1_Y. It follows from Proposition 4.10(c) that $M \supset 1_X$. Because M is minimal, it must equal 1_X. Thus, π satisfied condition (v) of Theorem 4.5(c) and so by (v)$\Rightarrow$(i) of Theorem 4.5(c), π is proximal. $\square$

The special properties of irreducible action maps of classic dynamical systems were studied by Auslander and Glasner (1977) where they are called *highly proximal* maps.

CHAPTER 5

Classical Dynamical Systems

Most familiar dynamical systems can be regarded as actions of one of the monoids $\mathbb{Z}_+$ or $\mathbb{R}_+$ or of a a topological group like $\mathbb{Z}$ or $\mathbb{R}$. The Ellis actions are obtained by compactifying via the Stone-Čech compactification or the enveloping semigroup. In these cases we have some important extra structure.

We begin with a bit of topology. If A is a not necessarily compact, Hausdorff space, then $B \subset A$ is called *bounded in* A if its closure in A is compact, otherwise it is called *unbounded in* A. Recall that the space A is called *separable* when it admits a countable dense subset.

Lemma 5.1. Let S be a compact space, G be a dense subset of S and U be an open subset of S.

(a) $U \cap G$ is dense in U.
(b) If $U \cap G$ is bounded in G then $U \subset G$.
(c) G is open in S iff it is locally compact.
(d) If G is open in S then it is separable iff S is separable.

PROOF. (a): If a nonempty $V \subset U$ is open with respect to U then it is open with respect to S and so meets the dense set G. Since $V \subset U$, it meets $U \cap G$.

(b): If $U \cap G \subset K$ and K is a compact subset of G then K is closed in S. By (a), $U \subset K \subset G$.

(c): Any open subset of a locally compact space, and so of a compact space, is locally compact. If G is locally compact and $x \in G$ then there is an open set V in S such that $V \cap G$ is a G neighborhood of x which is bounded in G. By (b) $V \subset G$. As G is an S neighborhood of each of its points it is open in S.

(d): If D is a countable dense subset of G then D is dense in S because G is dense in S. On the other hand, if D is a countable dense subset of S then because G is open $D \cap G$ is dense in G by (a). □

Definition 5.2. A *classical Ellis semigroup* is a triple (S, G, S^*) where

- S is an Ellis monoid with identity element e.

- G is an open, dense submonoid of S such that the restriction of the multiplication map

(5.1) $$M : G \times S \to S$$

is continuous. In particular, G is a locally compact, topological monoid.
- S^* is a nonempty, closed subset of S such that

(5.2) $$G \cup S^* = S.$$

and

(5.3) $$S \cdot S^* \cdot S = S^*.$$

In particular, S^* is a closed ideal in S.

A classical Ellis semigroup, (S, G, S^*), is called *separable* when G is separable, or, equivalently, when S is separable.

REMARK 5.1. Usually S^* is a proper closed ideal in S, but we can always obtain another example by using $S^* = S$ instead.

Definition 5.3. Let (S, G, S^*) be a classical Ellis semigroup and let X be a space, i.e. a compact Hausdorff space. A *classical action* $\Phi : S \times X \to X$ is an Ellis monoid action such that the restriction

$$\Phi : G \times X \to X$$

is continuous.

REMARK 5.2. Traditionally a G action $\Phi : G \times X \to X$ is denoted by (X, G, Φ), or just (X, G). Note that for any $x \in X$, $Sx = \overline{Gx}$ is the *orbit closure* of x under the G-action. For a $\mathbb{Z}_+$ action with $S^* = \beta\mathbb{Z}_+ \setminus \mathbb{Z}_+$ we have $S^*x = \omega(x)$.

If I is a directed index set, then an *inverse system* indexed by I is a family of Ellis actions $\{\Phi_\alpha : S \times X_\alpha \to X_\alpha\}$ together with a continuous action maps $\{\pi_{\beta\alpha} : X_\alpha \to X_\beta\}$ defined when α follows β in I and which satisfy $\pi_{\beta\alpha} \circ \pi_{\alpha\gamma} = \pi_{\beta\gamma}$. The associated *inverse limit* X is the closed subsystem of the product ΠX_α consisting of the points x such that $\pi_{\beta\alpha}(x_\alpha) = x_\beta$. We call the system a *surjective inverse system* when all of the $\pi_{\beta\alpha}$'s are surjective and we then call the limit a *surjective inverse limit*. By compactness each projection map $\pi_\alpha : X \to X_\alpha$ is then surjective.

We collect the following results which are mostly obvious consequences of the definitions. Recall from Definition 3.9 that an Ellis action Φ is called *densely continuous* when the maps Φ^p are continuous for a dense set of elements p.

Proposition 5.4. Let (S, G, S^*) be a classical Ellis semigroup.

(a) The translation action on S and any trivial action are classical Ellis actions.
(b) A classical Ellis action is densely continuous.
(c) Any subsystem of a classical Ellis action is a classical Ellis action. That is, if $\Phi : S \times X \to X$ is a classical Ellis action and K is a closed invariant subset of X then the restriction $\Phi|K : S \times K \to K$ is a classical Ellis action.
(d) Any factor of a classical Ellis action is a classical Ellis action. That is, if $\Phi : S \times X \to X$ is a classical Ellis action, $\Psi : S \times Y \to Y$ is an Ellis action and $\pi : X \to Y$ be a surjective continuous action map, then Ψ is a classical Ellis action.
(e) Any product or inverse limit of classical Ellis actions is a classical Ellis action.

PROOF. (a), (b) and (c) are obvious from the definitions.

(d): Observe that if R is any compact subset of G then just as in the proof of Proposition 3.3(c), $\Psi \circ (1_S \times \pi) = \pi \circ \Phi$ implies that on the compact space $R \times Y$ the restriction of Ψ is continuous. Since G is locally compact, it is the union of compact subsets R whose interiors cover G. So it follows that Ψ is continuous on $G \times Y$.

(e): If $X = \Pi_\alpha X_\alpha$ then the composition $\pi_\alpha \circ \Phi = \Phi_\alpha \circ (1_G \times \pi_\alpha) : G \times X \to X_\alpha$ is continuous for each α and so Φ is continuous. An inverse limit is a subsystem of a product and so is continuous by (c). □

Theorem 5.5. A classical action of a separable classical Ellis semigroup (S, G, S^*) is isomorphic to a surjective inverse limit of classical actions on metrizable spaces.

PROOF. Let $\Phi : S \times X \to X$ be a classical Ellis action and let G_0 be a countable dense monoid of G. Let $\mathcal{C}(X)$ denote the Banach algebra of real-valued continuous maps on X. For $p \in S$ and $f \in \mathcal{C}(X)$ let

(5.4) $\qquad p^* f \quad = \quad f \circ \Phi^p \qquad \text{so that} \qquad p^* f(x) \quad = \quad f(px).$

So if $p \in G$, Φ^p continuous implies that $p^* : \mathcal{C}(X) \to \mathcal{C}(X)$ is an algebra homomorphism of norm 1. In fact, uniform continuity implies that the associated map $\Phi^* : G \times \mathcal{C}(X) \to \mathcal{C}(X)$ is a topological action of the monoid G^{opp} with the order of multiplication reversed. We now proceed with a version of the classical Gelfand Construction (see, e.g. Akin (1997) Chapter 5). Let $\mathcal{F}$ be the collection of countable, G_0 invariant, subsets of the unit ball in $\mathcal{C}(X)$. That is, if $F \in \mathcal{F}$ then F is a countable collection of continuous maps from X to $I = [-1, 1]$ such that

(5.5) $\qquad f \in F \quad \text{and} \quad g \in G_0 \quad \implies \quad g^* f \in F.$

Each $F \in \mathcal{F}$ generates a separable subalgebra of $\mathcal{C}(X)$ and so to each there is a naturally associated metrizable compact space X_F and a continuous surjective map $\pi_F : X \to X_F$. Explicitly, we can use the image of the product space map:

(5.6)
$$\begin{aligned} \pi_F : X \to I^F \quad &\text{by} \quad (\pi_F(x))_f = f(x) \\ \text{so that} \quad \pi_f \circ \pi_F &= f \\ \text{and let} \quad X_F &=_{def} \pi_F(X). \end{aligned}$$

Since F is countable, X_F is a metrizable, compact space.

Because the algebra is invariant with respect to the action of the elements of G_0 we obtain an action of the semigroup G_0 (we can ignore the topology for the moment) on X_F such that the map π_F is an action map of G_0 actions (e.g. see Proposition 5.12 of Akin(1997)). It then follows from Proposition 3.11 that this action extends to an action of S on X_F which is the unique action such that π_F is an action map of S actions. Furthermore, this action is an Ellis action. Then by Proposition 5.4 the action is a classical Ellis action. Finally, $\mathcal{F}$ is directed with respect to the inclusion ordering, since $F_1, F_2 \in \mathcal{F}$ implies $F_1 \cup F_2 \in \mathcal{F}$. Given two points in X they can be distinguished by some set $F \in \mathcal{F}$ and so X is isomorphic to the inverse limit of the X_F's. $\square$

Most motivating examples use the Stone-Čech compactification construction.

Example (1) If G is a locally compact, topological group then the neighborhoods of identity, e, define the *right-invariant uniformity*, $\mathcal{U}_G$, by associating to the neighborhood U of e the relation

(5.7) $\qquad V_U \quad =_{def} \quad \{(a, b) \in G \times G : ab^{-1} \in U\}.$

We let $\mathcal{B}^u(G)$ denote the Banach algebra of bounded, real-valued functions which are uniformly continuous with respect to $\mathcal{U}_G$. Associated with this Banach algebra is the uniform Stone-Čech compactification $\beta_u G$. Each element $f \in \mathcal{B}^u(G)$ extends uniquely to a continuous real-valued function $\hat{f}$ on $\beta_u G$. Furthermore, the topological translation action of G on itself extends to a topological action $G \times \beta_u G \to \beta_u G$ (e.g. follow the proof of Proposition 5.12 of Akin (1997)). Because G is locally compact it is identified with an open dense subset of $\beta_u G$.

Because the uniformity is right invariant, each $p \in \beta_u G$ defines an algebra homomorphism $p^* : \mathcal{B}^u(G) \to \mathcal{B}^u(G)$ by

(5.8) $\qquad p^* f(a) \quad =_{def} \quad \hat{f}(ap) \quad \text{for} \quad a \in G.$

Hence, each right translation extends to a continuous map on $\beta_u G$. This defines the monoid multiplication on $\beta_u G$ which extends the group multiplication on G. Because G is a group we have

(5.9) $\quad g_1, g_2 \in G, \quad p \in \beta_u G, \quad$ and $\quad g_1 p g_2 \in G \implies p \in G.$

Finally, we denote by

(5.10) $\quad\quad\quad \beta_u^* G \quad =_{def} \quad \beta_u G \setminus G.$

when G is a proper subset of $\beta_u G$, i.e. when G is not compact. When G is compact then $G = \beta_u G$ and we define $\beta_u^* G = G$ as well in this case. It follows that $(S, G, S^*) = (\beta_u G, G, \beta_u^* G)$ is a classical Ellis semigroup which is separable iff G is. From (5.9) and continuity of right translation it follows that condition (5.3) holds. To be precise, the contrapositive of (5.9) says $GS^*G \subset S^*$ and so continuity implies $SS^* \subset S^*$. Also, $S^*S = S^*(G \cup S^*) \subset S^*$.

A topological action of G on a space X extends to a classical Ellis action of this classical semigroup on X.

Example (2) The group of real numbers, $\mathbb{R}$, is an abelian, separable, locally compact topological group and so is a special case of Example (1). A topological action of $\mathbb{R}$ on a space X is a *flow* on X. A *semiflow* on X is a topological action of the monoid $\mathbb{R}_+$ on X. We denote by $\beta_u \mathbb{R}_+$ the closure in $\beta_u \mathbb{R}$ of $\mathbb{R}_+$ and let

(5.11) $\quad\quad\quad \beta_u^* \mathbb{R}_+ \quad =_{def} \quad \beta_u \mathbb{R}_+ \setminus \mathbb{R}_+ \quad = \quad (\beta_u \mathbb{R}_+) \cap (\beta_u^* \mathbb{R}).$

Because translation by a negative real number $-t$ maps the complement of the compact set $[0, t]$ into $\mathbb{R}_+$, the translation on $\beta_u \mathbb{R}$ maps $\beta_u^* \mathbb{R}_+$ into itself. Thus, $\beta_u^* \mathbb{R}_+$ is a closed ideal in $\beta_u \mathbb{R}$. Condition (5.3) for $\beta_u \mathbb{R}_+$ follows from the corresponding condition for $\beta_u \mathbb{R}$. A semiflow on X extends to a classical action of the separable, classical Ellis semigroup $(\beta_u \mathbb{R}_+, \mathbb{R}_+, \beta_u^* \mathbb{R}_+)$.

Example (3) A discrete group is a locally compact topological group and $\mathcal{U}_G$ is the discrete uniformity containing the diagonal 1_G. The discrete topology is separable iff the group is countable. In this case, we have the ordinary Stone-Čech compactification. The classical Ellis semigroup is $(\beta G, G, \beta^* G)$. Any action of G on a space extends to a classical action of $(\beta G, G, \beta^* G)$.

Example (4) The group of integers, $\mathbb{Z}$, is a countable, abelian, discrete group and so is a special case of Example (3). An action Φ of $\mathbb{Z}$ on a space X consists of the positive and negative iterates of a homeomorphism f on X with $\Phi^1 = f$. Positive time iteration of a continuous map $f : X \to X$ is a monoid action of $\mathbb{Z}_+$ on X. The Stone-Čech compactification of $\mathbb{Z}_+$ is the closure in $\beta \mathbb{Z}$ of $\mathbb{Z}_+$. As before

(5.12) $\quad\quad\quad \beta^* \mathbb{Z}_+ \quad =_{def} \quad \beta \mathbb{Z}_+ \setminus \mathbb{Z}_+ \quad = \quad (\beta \mathbb{Z}_+) \cap (\beta^* \mathbb{Z}).$

As in Example (2), $\beta^* \mathbb{Z}_+$ is a closed ideal in $\beta \mathbb{Z}$.

Iteration of the continuous map f extends to a classical action of the separable, classical Ellis semigroup $(\beta \mathbb{Z}_+, \mathbb{Z}_+, \beta^* \mathbb{Z}_+)$. In this important special case, the set $\omega f(x)$, the set of limit points of the orbit sequence $\{f(x), f^2(x), ...\}$ is given by

(5.13) $$\omega f(x) = \beta^* \mathbb{Z}_+ x.$$

A related pair of examples is the *one-point compactification* of $\mathbb{Z}_+$ and of $\mathbb{Z}$ which are the quotients obtained by smashing β^* to a point denoted ∞. We obtain the classical semigroups $(\mathbb{Z}_+ \cup \{\infty\}, \mathbb{Z}_+, \{\infty\})$ and $(\mathbb{Z} \cup \{\infty\}, \mathbb{Z}, \{\infty\})$

A classical action $\Phi : S \times X \to X$ of a classical semigroup (S, G, S^*) restricts to a topological action of the locally compact monoid G on X and to an Ellis action of the closed ideal S^*. Before proceeding we sort out how the different associated notions of invariance, distality, etc. are related among these actions. We say that a subset $A \subset X$ is G *minus invariant* when

(5.14) $\quad x \in X, \quad g \in G \quad \text{and} \quad gx \in A \quad \Longrightarrow \quad x \in A.$

This condition says exactly that the complement $X \setminus A$ is G invariant. Compare it with the capturing property for S.

Proposition 5.6. Let (S, G, S^*) be a classical Ellis semigroup, let $\Phi : S \times X \to X$ be a classical Ellis action and let A be a subset of X.

(a) The following are equivalent:
  (i) A is S invariant.
  (ii) A is $\Phi^\#(S)$ invariant.
  (iii) A is G invariant and S^* invariant.
  (iv) A is G invariant and $\Phi^\#(S^*)$ invariant.
(b) If A is closed and G invariant then it is S invariant.
(c) If A is G invariant then its closure is G invariant and so is S invariant.
(d) If $A = S^* B$ for some subset B of X then A is S invariant. In particular, if $A = S^* A$ then A is S invariant.
(e) If A is open and G minus invariant then A satisfies the capturing property

(5.15) $\quad x \in X, \quad p \in S \quad \text{and} \quad px \in A \quad \Longrightarrow \quad x \in A.$

PROOF. (a): Invariance with respect to an acting Ellis semigroup and the associated enveloping semigroup are always the same. (i) $\Leftrightarrow$ (iii) because $S = G \cup S^*$.

(b): If A is closed and $x \in A$ then continuity of Φ_x implies that $(\Phi_x)^{-1}(A) = \{p : px \in A\}$ is closed in S. If it contains the dense set G then it is all of S.

(c): Lemma 3.10(b) and (b) above.
(d): S^* is an ideal.
(e): Apply (b) to the complement of A. □

REMARK 5.3. On the triangle $X = \{(t,s) : 0 \leq t \leq s \leq 1\} \subset \mathbb{R}^2$ define a classical $(S, G, S^*) = (\beta \mathbb{Z}_+, \mathbb{Z}_+, \beta^* \mathbb{Z}_+)$ action by $\Phi^1 = f : X \to X$ given by $f(t, s) = (t^2 s, s^2)$. Let Y be the segment $\{(t, 1) : 0 \leq t \leq 1\}$. For every $x \in X \setminus Y$ the limit set $\omega f(x)$ is the fixed point $(0, 0)$. Thus, for all $p \in S^*$, $px = (0, 0)$ if $x \in X \setminus Y$. On the other hand, for all $p \in S^*$, $px = (0, 1)$ if $x \in Y \setminus (1, 1)$. Thus, the set $A = \{(s/2, s) : 0 \leq s < 1\}$ is S^* invariant but its closure is not. The trouble is that the action of $S^* = \beta^* \mathbb{Z}_+$ is not densely continuous although, of course, the action of $S = \beta \mathbb{Z}_+$ is.

Finally, the closed set $\{(0, 0), (1/4, 1/2)\}$ is S^* invariant but not $\mathbb{Z}_+$ invariant.

Corollary 5.7. Let (S, G, S^*) be a classical Ellis semigroup.
(a) $[Min(S)]$ is the closure of the ideal $Min(S)$ of minimal points in S and $[Min(S)] \subset S^*$.
(b) Let $\Phi : S \times X \to X$ be a classical Ellis action. The set $Min(X)$ of minimal points is S invariant and its closure is the min-center $[Min(X)]$. A point is minimal iff it is minimal with respect to the restricted action $\Phi|S^*$.

PROOF. (a): Apply Lemma 3.10(b) and Lemma 1.1(b). (b): Apply Lemma 3.10(b) and Lemma 4.1(c). □

Since S^* contains all the minimal points of S it follows from Proposition 2.4(c) that for a classical Ellis action of (S, G, S^*)

(5.16) $\quad\quad\quad PROX \quad = \quad PROX_{S^*}.$

Hence, the notions of proximality and distality agree for the S action and the restricted S^* action. Since S is a monoid, every point is S recurrent and only diagonal pairs are S asymptotic. Distality, almost distality and semidistality for the S action collapse together to distality which agrees with distality for the restricted S^* action. Hence, we adopt the following:

Convention For a classical Ellis action $\Phi : S \times X \to X$ of a classical Ellis semigroup (S, G, S^*)

- We will use *invariance* to mean S invariance. For G invariance or S^* invariance we will use explicit labels. Of course, if the subset is closed and G invariant then it is automatically invariant by Proposition 5.6(b).
- We will use *recurrence* to mean S^* recurrence and *asymptotic* to mean S^* asymptotic. So we say that the classical action Φ is distal,

almost distal or semidistal when the restricted S^* action is distal, almost distal or semidistal.

Thus, a point x is recurrent iff Iso_x, which always contains the identity, meets S^*.

Definition 5.8. For a classical Ellis action $\Phi : S \times X \to X$ of a classical Ellis semigroup (S, G, S^*), we will say that *the G action is surjective* when Φ^g is surjective on X for every $g \in G$.

We say that a classical Ellis semigroup (S, G, S^*) satisfies the *Surjection Condition* when

(5.17) $\qquad M^g(S^*) \;=\; S^* \;=\; M_g(S^*) \qquad$ for all $\;g \in G$.

REMARK 5.4. The Surjection Condition says exactly that the left and right G actions on S^* are surjective. Note that the left action is the restriction of the translation action to S^*. Examples (1)-(4) all satisfy the Surjection Condition. Of course, when G is a group all G actions are surjective.

Proposition 5.9. Let (S, G, S^*) be a classical Ellis semigroup and let $\Phi : S \times X \to X$ be a classical Ellis action. $ASYMP$ is an invariant, i.e. S invariant, subset of $X \times X$.

If (S, G, S^*) satisfies the Surjection Condition then $RECUR$ is a G invariant subset of X and $PROX$ is a G invariant subset of $X \times X$. In that case, the center $[RECUR]$ is the closure of the set of recurrent points.

PROOF. If $(x_1, x_2) \in ASYMP$ and $p \in S$ then for all $q \in S^*$, $qp \in S^*$ implies $qpx_1 = qpx_2$ and so $(px_1, px_2) \in ASYMP$ (see Proposition 1.10(b)).

Now assume the Surjection Condition. If $x \in RECUR$ and $g \in G$ then

(5.18) $\qquad M^g(M_g^{-1}(Iso_x)) \;\subset\; Iso_{gx}.$

That is, if $pg \in Iso_x$ then $gp \in Iso_{gx}$. If $Iso_x \cap S^* \neq \emptyset$ then the Surjection Condition implies that $Iso_{gx} \cap S^* \neq \emptyset$. Similarly, for $x_1, x_2 \in X$

(5.19) $\qquad M_g^{-1}(Foc_{(x_1,x_2)}) \;\subset\; Foc_{(gx_1,gx_2)}$

implies that $PROX$ is G invariant.

By Proposition 5.6(c) the closure of $RECUR$ is invariant and so is the smallest closed, invariant set containing $RECUR$. $\qquad\square$

Corollary 5.10. Let (S, G, S^*) be a classical Ellis semigroup which satisfies the Surjection Condition. Let $\Phi : S \times X \to X$ be a classical Ellis action.

(a) $PROX$ is a closed subset of $X \times X$ iff the only minimal pairs in its closure lie in the diagonal, i.e. $\overline{PROX} = PROX$ iff

(5.20) $$\overline{PROX} \cap Min(X \times X) \subset 1_X.$$

(b) Assume that $PROX$ is a closed subset of $X \times X$. $PROX$ is then an invariant, closed, equivalence relation on X and Φ is a $[Min(S)]$ almost distal system.

Let Y denote the space of $PROX$ equivalence classes with quotient topology from the associated projection map $\pi : X \to Y$. There is a unique action $\Psi : S \times Y \to Y$ such that π is an action map. Furthermore, Ψ is a distal classical Ellis action.

PROOF. (a): Since a minimal subset of $X \times X$ is either contained in the diagonal or is disjoint from the diagonal, we have, for any semigroup action:

(5.21) $$PROX \cap Min(X \times X) \subset 1_X.$$

So (5.20) is obvious when $PROX$ is closed. Now assume (5.20). By Proposition 5.9 $PROX$ is G invariant and so $\overline{PROX}$ is invariant, i.e. S invariant, by Proposition 5.6(c). If $z \in \overline{PROX}$ and $p \in Min(S)$ then $pz \in \overline{PROX} \cap Min(X \times X)$. By (5.20) this is a diagonal pair. Thus, $p \in Foc_z$ and so $z \in PROX$.

(b): As it is closed and G invariant, $PROX$ is invariant. By Theorem 4.6(a) Φ is $[Min(S)]$ almost distal and $PROX$ is an equivalence relation.

Because the equivalence relation is invariant, the action Ψ is well-defined by $pPROX(x) = PROX(px)$. It is an Ellis action by Lemma 3.11 and so is a classical Ellis action by Proposition 5.4(d). The projection map is obviously proximal and so by Lemma 1.6(c), $(\pi \times \pi)^{-1}(PROX(\Psi)) = PROX(\Phi) = R_\pi$. Hence, $PROX(\Psi) = 1_Y$. That is, the quotient system is distal. □

REMARK 5.5. Conversely, if Φ is a proximal extension by an action map π of a distal system Ψ then $PROX(\Phi) = R_\pi$ and so is closed. Furthermore, Ψ is isomorphic to the quotient system associated with the equivalence relation $PROX(\Phi)$.

In our further study of classical dynamical systems we will use two conceptual schemes which have proved useful for the dynamics of map iterations. First, we recall the language of Furstenberg families, as in Furstenberg (1981). We follow the notation of Akin (1997) Chapter 1.

For a set G a *family on G* is a collection of subsets $\mathcal{F}$ of G which is *hereditary upwards*. That is, $A \in \mathcal{F}$ and $A \subset B$ imply $B \in \mathcal{F}$. A family is called *proper* when it is a proper subset of the power set $\mathcal{P}$ of G. Clearly, a family $\mathcal{F}$ is proper iff $G \in \mathcal{F}$ and $\emptyset \notin \mathcal{F}$. For any family $\mathcal{F}$ the *dual* $k\mathcal{F}$ is given by:

(5.22) $$\begin{aligned} k\mathcal{F} \quad =_{def} \quad & \{A \in \mathcal{P} : A \cap B \neq \emptyset \quad \text{for all} \quad B \in \mathcal{F}\} \\ = \quad & \{A \in \mathcal{P} : G \setminus A \notin \mathcal{F}\}. \end{aligned}$$

That is, $G \setminus A \in \mathcal{F}$ iff $A \cap B = \emptyset$ for some $B \in \mathcal{F}$.

It is easy to check that $kk\mathcal{F} = \mathcal{F}$ and that the operator k reverses inclusions between families. The two improper families are related by $k\mathcal{P} = \emptyset$. If $\mathcal{A} \subset \mathcal{P}$ then the *family generated by* $\mathcal{A}$ is

(5.23) $$[[\mathcal{A}]] \quad =_{def} \quad \{A \in \mathcal{P} : B \subset A \quad \text{for some} \quad B \in \mathcal{A}\}.$$

Clearly,

(5.24) $$k[[\mathcal{A}]] \quad = \quad \{A \in \mathcal{P} : A \cap B \neq \emptyset \quad \text{for all} \quad B \in \mathcal{A}\}$$

If $\mathcal{F}_1$ and $\mathcal{F}_2$ are two families then their *join* is

(5.25) $$\mathcal{F}_1 \cdot \mathcal{F}_2 \quad =_{def} \quad \{A_1 \cap A_2 : A_1 \in \mathcal{F}_1, A_2 \in \mathcal{F}_2\}.$$

If $\mathcal{F}_1$ and $\mathcal{F}_2$ are proper, it is easy to see that $\mathcal{F}_1 \cdot \mathcal{F}_2$ is proper iff

(5.26) $$\mathcal{F}_1 \subset k\mathcal{F}_2.$$

A family $\mathcal{F}$ is a *filter* when it is proper and it is closed under finite intersection or, equivalently, when $\mathcal{F} \cdot \mathcal{F} = \mathcal{F}$. The dual of a filter is called a *filterdual*. A proper family is a filterdual exactly when it satisfies what Furstenberg calls the *Ramsey Property* namely, $A_1 \cup A_2 \in \mathcal{F}$ implies $A_1 \in \mathcal{F}$ or $A_2 \in \mathcal{F}$. In Glasner (1980) it was called a *divisible property*.

If $G \subset S$ and $\mathcal{T}$ is a family on S then the *trace of* $\mathcal{T}$ *on* G is

(5.27) $$\mathcal{T} \wedge G \quad =_{def} \quad \{A \cap G : A \in \mathcal{T}\}.$$

The trace is proper iff $S \setminus G \notin \mathcal{T}$. For (S, G, S^*) a classical Ellis semigroup, the most important families are the proper families on G given by:

(5.28) $$\begin{aligned} \mathcal{P}_+ \quad =_{def} \quad & \{A \subset G : A \neq \emptyset\}. \\ k\mathcal{P}_+ \quad = \quad & \{G\}. \\ \mathcal{B} \quad =_{def} \quad & \{A \subset G : \overline{A} \cap S^* \neq \emptyset\}. \\ k\mathcal{B} \quad = \quad & \{A \subset G : S^* \subset Int(A \cup (S \setminus G))\}. \end{aligned}$$

Thus, $\mathcal{P}_+$ is the entire power set $\mathcal{P}$ except for the empty set. Its dual is the singleton filter $\{G\}$. They are, respectively, the largest and smallest proper families on G. Usually, $\mathcal{B}$ and $k\mathcal{B}$ are, respectively, the families of unbounded subsets and cobounded subsets of G. For example, when $G = \mathbb{Z}_+$ (as in Example (4)), a set $A \subset \mathbb{Z}_+$ is in $\mathcal{B}$ when it is infinite and is in $k\mathcal{B}$ when it is cofinite.

In detail, for $\mathcal{B}$ and its dual the closure and interior operators are taken in S. $k\mathcal{B}$ is the trace on G of the filter on S of neighborhoods of S^*. Since G is dense in S every nonempty open subset of S meets G and so $k\mathcal{B}$ is a filter on G. Observe that if $A \subset G$ then its closure in S is disjoint from S^* exactly when its complement in S, i.e. $S \setminus A = (G \setminus A) \cup (S \setminus G)$ is a neighborhood of S^*. Thus, as described in (5.28) $\mathcal{B}$ and $k\mathcal{B}$ are dual families.

Notice that $A \subset G$ is unbounded in G exactly when its closure in S meets $S \setminus G$. Thus, every unbounded subset of G lies in $\mathcal{B}$. In the usual case that the inclusion $S \setminus G \subset S^*$ is an equality, the filterdual $\mathcal{B}$ consists exactly of the unbounded subsets of G. At the other extreme, when $S^* = S$, $\mathcal{B} = \mathcal{P}_+$.

If (S, G, S^*) is a classical Ellis semigroup and $\mathcal{F}$ is a filter of subsets of G, then we will define the *hull of* $\mathcal{F}$ to be

$$(5.29) \qquad H(\mathcal{F}) \quad =_{def} \quad \bigcap_{F \in \mathcal{F}} \overline{F} \quad \subset \quad S,$$

where the closure is taken in S. For example,

$$(5.30) \qquad H(k\mathcal{P}_+) \quad = \quad S \quad \text{and} \quad H(k\mathcal{B}) \quad = \quad S^*.$$

For a classical Ellis action $\Phi : S \times X \to X$ of (S, G, S^*) the family language is mostly used to provide information about the *hitting time sets*. If $A, B \subset X$ then we let

$$(5.31) \qquad \begin{aligned} N^\Phi(A, B) \quad &=_{def} \quad \{g \in G : \Phi^g(A) \cap B \neq \emptyset\} \\ &= \quad \{g \in G : A \cap (\Phi^g)^{-1}(B) \neq \emptyset\}. \end{aligned}$$

We will often omit the superscript Φ when the action is understood. We will write $N^\Phi(x, B)$ for $N^\Phi(\{x\}, B)$ when $x \in X$. For a family $\mathcal{F}$ various properties will be expressed by considering when $N^\Phi(U, V)$ or $N^\Phi(x, V)$ lies in $\mathcal{F}$ for certain opene $U, V \subset X$. Recall that an *opene* set is an open, nonempty set. Notice that when V is open in G, then and $N^\Phi(x, V)$ is open in G.

We also need to briefly review the ideas associated with a relation on a set. Recall that, as a matter of set theory, a function $f : X \to Y$ is a subset of $X \times Y$, namely the set of pairs $(x, f(x))$ as x varies over X. For example, the identity map 1_X is the diagonal in $X \times X$. Conversely, any $R \subset X \times Y$ can be regarded as a *relation* from X to Y. Then for $x \in X$ or $A \subset X$, let

$$(5.32) \qquad \begin{aligned} R(x) \quad &=_{def} \quad \{y \in Y : (x, y) \in R\}. \\ R(A) \quad &=_{def} \quad \bigcup_{x \in A} R(x) \quad = \quad p_2(R \cap (A \times Y)) \\ &= \quad \{y \in Y : (x, y) \in R \quad \text{for some} \quad x \in A\}, \end{aligned}$$

where $p_2 : X \times Y \to Y$ is the projection map. The inverse relation from Y to X is

$$(5.33) \qquad R^{-1} \quad =_{def} \quad \{(y, x) : (x, y) \in R\}.$$

Observe that if R is a function then $R(A)$ and $R^{-1}(B)$ are just the usual image of A and pre-image of $B \subset Y$, respectively.

If S is a relation from Y to Z, i.e. a subset of $Y \times Z$, then we define the composition

(5.34)
$$\begin{aligned} S \circ R &=_{def} \{(x,z) : (x,y) \in R, (y,z) \in S \quad \text{for some} \quad y \in Y\} \\ &= p_{13}((R \times Z) \cap (X \times S)), \end{aligned}$$

where $p_{13} : X \times Y \times Z \to X \times Z$ is the projection map.

R is a *closed relation* from X to Y when it is a closed subset of $X \times Y$. Since our spaces are compact, a function from X to Y is continuous precisely when, as a relation, it is closed. Closed relations between compact spaces have some of the nice properties of continuous maps. The elements of the theory are presented in Chapter 1 of Akin (1993). The image of a closed set under a closed relation is closed, the inverse of a closed relation is closed and the composition of closed relations is closed. It follows that if $U \subset Y$ is open and $R \subset X \times Y$ is closed then

(5.35)
$$\{x \in X : R(x) \subset U\} = X \setminus R(Y \setminus U)$$

is open.

For any Ellis action $\Phi : S \times X \to X$ there is an associated closed relation. Notice that, as it is a function, $\Phi \subset S \times X \times X$.

(5.36)
$$\mathcal{N}\Phi =_{def} \overline{p_{23}(\Phi)} = p_{23}(\overline{\Phi}),$$

where $p_{23} : S \times X \times X \to X \times X$ is the projection map. $\mathcal{N}\Phi$ is called the *prolongation* of the action Φ (see Auslander (1964)). It is clear that for all $x \in X$, the prolongation set $\mathcal{N}\Phi(x)$ contains the orbit Sx which equals $H(k\mathcal{P}_+ x)$ by (5.30). Corollary 5.14 below will show that under some circumstances equality holds for all x in a residual subset of X.

First we define the prolongation associated with any family $\mathcal{F}$ of subsets of G. For a classical action we define

(5.37)
$$\mathcal{N}_{\mathcal{F}}\Phi =_{def} \bigcap_{F \in k\mathcal{F}} \overline{\pi_{23}(\Phi \cap (F \times X \times X))} = \bigcap_{F \in k\mathcal{F}} \overline{\bigcup_{g \in F} \Phi^g}.$$

Lemma 5.11. Let (S, G, S^*) be a classical Ellis semigroup, $\mathcal{F}$ be a family of subsets of G and $\Phi : S \times X \to X$ be a classical Ellis action. Assume that $\mathcal{T}$ is an opene basis for the topology of X.

(a) $\mathcal{N}\Phi = \mathcal{N}_{\mathcal{P}_+}\Phi$.
(b) For points $x, y \in X$ we have $y \in \mathcal{N}_{\mathcal{F}}\Phi(x)$, i.e. $(x,y) \in \mathcal{N}_{\mathcal{F}}\Phi$ iff $N^{\Phi}(U,V) \in \mathcal{F}$ for every pair of open sets $U, V \subset X$ such that $x \in U$ and $y \in V$.
(c) Assume that $k\mathcal{F}$ is a filter. For $x \in X$ we have

(5.38)
$$H(k\mathcal{F})x \subset \mathcal{N}_{\mathcal{F}}\Phi(x).$$

On the other hand, $y \notin \mathcal{N}_{\mathcal{F}}\Phi(x)$ iff for some $V \in \mathcal{T}$ with $y \in V$ and some $F \in k\mathcal{F}$, $x \notin \overline{\bigcup_{g \in F} (\Phi^g)^{-1}(V)}$.

(d) Assume that $k\mathcal{F}$ is a filter generated as a family by $\mathcal{A} \subset k\mathcal{F}$.

$$
(5.39) \quad \begin{aligned} \{x \in X : \mathcal{N}_{\mathcal{F}}\Phi(x) = H(k\mathcal{F})x\} \quad &= \\ \bigcap\{X \setminus Bdry[\bigcup_{g \in F}(\Phi^g)^{-1}(V)] &: F \in \mathcal{A}, V \in \mathcal{T}\}. \end{aligned}
$$

where $Bdry$ denotes the topological boundary, i.e. for an open subset U of X, $\quad X \setminus Bdry[U] = U \cup (X \setminus \overline{U})$.

(e) Let $\Psi : S \times Y \to Y$ be a classical Ellis action and $\pi : X \to Y$ be a surjective action map.

$$
(5.40) \quad (\pi \times \pi)(\mathcal{N}_{\mathcal{F}}\Phi) \quad \subset \quad \mathcal{N}_{\mathcal{F}}\Psi
$$

with equality when $k\mathcal{F}$ is a filter.

PROOF. (a): For a classical action, the restriction of Φ to $G \times X$, i.e. $\Phi \cap (G \times X \times X)$ is dense in Φ. Since $k\mathcal{P}_+ = \{G\}$ the equation is clear.

(b): Clearly, $(x,y) \in \mathcal{N}_{\mathcal{F}}\Phi$ iff for every pair of opens U, V with $(x,y) \in U \times V$, and every $F \in k\mathcal{F}$, $U \times V$ meets $\bigcup_{g \in F} \Phi^g$. So the result follows because

$$
(5.41) \quad (U \times V) \cap \Phi^g \quad \neq \quad \emptyset \quad \Longleftrightarrow \quad g \in N^{\Phi}(U,V),
$$

and $N^{\Phi}(U,V)$ meets every $F \in k\mathcal{F}$ exactly when it is a member of $\mathcal{F}$.

(c): First observe that when $k\mathcal{F}$ is a filter,

$$
(5.42) \quad \begin{aligned} H(k\mathcal{F})x \quad &= \quad \Phi_x(\bigcap_{F \in k\mathcal{F}} \overline{F}) \\ &= \quad \bigcap_{F \in k\mathcal{F}} \Phi_x(\overline{F}) \quad = \quad \bigcap_{F \in k\mathcal{F}} \overline{\{\Phi^g(x) : g \in F\}}. \end{aligned}
$$

It follows that $y \in H(k\mathcal{F})x$ iff $N^{\Phi}(x,V) \in \mathcal{F}$ for all open V such that $y \in V$. In particular, it follows from part (b) that for all x inclusion (5.38) holds.

Furthermore, $y \in H(k\mathcal{F})x$ iff $x \in \overline{\bigcup_{g \in F}(\Phi^g)^{-1}(V)}$ for all $F \in k\mathcal{F}$ and V open with $y \in V$. Clearly, it suffices that these conditions hold for all $F \in \mathcal{A}$ and $V \in \mathcal{T}$ with $y \in V$ when $\mathcal{A}$ generates $k\mathcal{F}$.

On the other hand, $y \notin \mathcal{N}_{\mathcal{F}}\Phi(x)$ iff for some $U, V \in \mathcal{T}$ with $(x,y) \in U \times V$, $N(U,V) \notin \mathcal{F}$ and so iff for some such U, V and some $F \in k\mathcal{F}$ $N(U,V) \cap F = \emptyset$, or, equivalently, $U \cap (\bigcup_{g \in F}(\Phi^g)^{-1}(V)) = \emptyset$. Hence, $y \notin \mathcal{N}_{\mathcal{F}}\Phi(x)$ iff for some $V \in \mathcal{T}$ with $y \in V$ and some $F \in k\mathcal{F}$, $\quad x \notin \overline{\bigcup_{g \in F}(\Phi^g)^{-1}(V)}$.

(d): It follows from (b) that if x is in the intersection given in (5.39) then for all $y \in X$ either $y \in H(k\mathcal{F})x$ or $y \notin \mathcal{N}_{\mathcal{F}}\Phi(x)$. Hence, $H(k\mathcal{F})x = \mathcal{N}_{\mathcal{F}}\Phi(x)$ for all x in the intersection.

Finally, if $y \notin H(k\mathcal{F})x$ then for some V and F as above $x \notin \bigcup_{g \in F}(\Phi^g)^{-1}(V)$. If also $y \in \mathcal{N}_{\mathcal{F}}\Phi(x)$ then for all such V and F, $\quad x \in \overline{\bigcup_{g \in F}(\Phi^g)^{-1}(V)}$. So for at least one $V \in \mathcal{T}$ and $F \in k\mathcal{A}$, x lies in the forbidden boundary.

(e): Because π is a surjective action map we have $(\pi \times \pi)(\Phi^g) = \Psi^g$ for all $g \in G$. Take the union over $g \in F$, take the closure and intersect over $F \in k\mathcal{F}$. The image of the intersection is contained in the intersection of the images with equality when the collection of compact sets being mapped is closed under finite intersection, i.e. when $k\mathcal{F}$ is a filter. $\square$

REMARK 5.6. We will denote $\mathcal{N}_\mathcal{B}\Phi$ by $\mathcal{N}^*\Phi$. Since $k\mathcal{P}_+$ and $k\mathcal{B}$ are filters:
$$(5.43) \quad (\pi \times \pi)(\mathcal{N}\Phi) = \mathcal{N}\Psi \quad \text{and} \quad (\pi \times \pi)(\mathcal{N}^*\Phi) = \mathcal{N}^*\Psi.$$
when $\pi : X \to Y$ is a surjective action map.

Clearly, $\mathcal{N}^*\Phi$ includes the prolongation of the restriction Φ_{S^*} of the Ellis action to S^*, but it might be strictly larger. For example, if the homeomorphism f on the unit interval defined by $f(t) = t^2$ is used to obtain a classical action Φ of $(\beta\mathbb{Z}_+, \mathbb{Z}_+, \beta^*\mathbb{Z}_+)$, then $\mathcal{N}\Phi_{\beta^*\mathbb{Z}_+}(1) = \{0,1\}$ but $\mathcal{N}^*\Phi(1)$ is the entire unit interval.

Proposition 5.12. Let (S, G, S^*) be a classical Ellis semigroup and let $\Phi : S \times X \to X$ be a classical Ellis action. For all $x \in X$, $\mathcal{N}\Phi(x)$ and $\mathcal{N}^*\Phi(x)$ are closed, invariant subsets of X. Furthermore, for all $x \in X$
$$(5.44) \quad \mathcal{N}\Phi(x) = Sx \cup \mathcal{N}^*\Phi(x).$$

PROOF. If $g \in G$ and $U, V \subset X$ then $h \in N(U, (\Phi^g)^{-1}(V))$ iff $gh \in N(U,V)$. Thus,
$$(5.45) \quad N^\Phi(U, (\Phi^g)^{-1}(V)) = (M^g)^{-1}(N^\Phi(U,V)).$$

Now assume that U and V are open with $x \in U$ and $gy \in V$. If $y \in \mathcal{N}\Phi(x)$ then $N^\Phi(U, (\Phi^g)^{-1}(V))$ is nonempty and so by equation (5.45) $N(U,V)$ is nonempty. It follows that $gy \in \mathcal{N}\Phi(x)$.

If A is any neighborhood of S^* in S then since $gS^* \subset S^*$, $(M^g)^{-1}(A)$ is a neighborhood of S^*. If $y \in \mathcal{N}^*\Phi(x)$ then $N^\Phi(U, (\Phi^g)^{-1}(V)) \cap (M^g)^{-1}(A)$ is nonempty and so by equation (5.45) $N(U,V) \cap A$ is nonempty. It follows that $gy \in \mathcal{N}^*\Phi(x)$.

Observe that from (5.42) and (5.30) we have for all $x \in X$
$$(5.46) \quad Sx \subset \mathcal{N}\Phi(x) \quad \text{and} \quad S^*x \subset \mathcal{N}^*\Phi(x).$$
It follows that $\mathcal{N}\Phi(x)$ contains the union on the right in (5.44). On the other hand, suppose that $y \in \mathcal{N}\Phi(x) \setminus \mathcal{N}^*\Phi(x)$. There exist open sets $U, V \subset X$ with $(x,y) \in U \times V$ and a compact $K \subset G \setminus S^*$ such that $N(U,V) \subset K$. By definition of $\mathcal{N}\Phi(x)$ there exist a net (x_i, g_i) in $X \times G$ such that x_i converges to x and $g_i x_i$ converges to y. By restricting to a cofinal subset of the index set we can assume that $(x_i, g_i x_i)$ all lie in $U \times V$ which implies $g_i \in K$. So by

going to a subnet we can assume g_i converges to an element $g \in K$. Because the action of G on X is topological, it follows that $g_i x_i$ converges to gx and so $gx = y$. That is, $y \in Sx$. □

Theorem 5.13. Let (S, G, S^*) be a classical Ellis semigroup, $\mathcal{F}$ be a family of subsets of G and $\Phi : S \times X \to X$ be a classical Ellis action. If $k\mathcal{F}$ is a countably generated filter and X is metrizable then $\{x \in X : \mathcal{N}_\mathcal{F}\Phi(x) = H(k\mathcal{F})x\}$ is a dense, G_δ subset of X.

PROOF. The boundary of an open set is a closed nowhere dense set. So if $k\mathcal{F}$ is countably generated and X has a countable base then the result follows from (5.38) in Lemma 5.11(c) and the Baire Category Theorem. □

The next corollary generalizes results in Akin (1993) and Akin-Glasner (1998).

Corollary 5.14. Let (S, G, S^*) be a classical Ellis semigroup and $\Phi : S \times X \to X$ be a classical Ellis action with X metrizable. $\{x : \mathcal{N}\Phi(x) = \overline{Sx}\} = \{x : \mathcal{N}\Phi(x) = \overline{Gx}\}$ is a dense, G_δ subset of X. If, in addition, S^* is a G_δ subset of S then $\{x : \mathcal{N}^*\Phi(x) = S^*x\}$ is a dense, G_δ subset of X.

PROOF. $k\mathcal{P}_+$ is a singleton and $k\mathcal{B}$ is countably generated when S^* is a G_δ. Apply Theorem 5.13 and equation (5.30). (See the remark following Definition 5.3.) □

For a closed relation R on a space X, the set

$$(5.47) \qquad |R| \quad =_{def} \quad \{x \in X : (x, x) \in R\} \quad = \quad \Delta^{-1}(R)$$

is a closed subset of X, where $\Delta : X \to X \times X$ is the diagonal map defined by $\Delta(x) = (x, x)$.

A point $x \in X$ is called *nonwandering* for a classical Ellis action Φ when $x \in |\mathcal{N}^*\Phi|$, i.e. when $x \in \mathcal{N}^*\Phi(x)$. The set $|\mathcal{N}^*\Phi|$ is called the *nonwandering set*.

Proposition 5.15. Let $\Phi : S \times X \to X$ be a classical Ellis action of a classical Ellis semigroup (S, G, S^*).

(a) A point $x \in X$ is nonwandering iff for every open $U \subset X$ with $x \in U$, $N^\Phi(U, U) \in \mathcal{B}$. If x is a recurrent point then it is nonwandering. In particular, every minimal point is nonwandering.

(b) If $RECUR$, the set of recurrent points of Φ, is dense in X then the nonwandering set $|\mathcal{N}^*\Phi|$ is equal to X. On the other hand, if G is separable and S^* is a G_δ subset of S then, conversely, $|\mathcal{N}^*\Phi| = X$ implies $RECUR$ is dense in X.

(c) If S^* is a G_δ subset of S, X is metrizable and $|\mathcal{N}^*\Phi| = X$ then $RECUR$ is a dense, G_δ subset of X.

PROOF. (a): The first sentence follows from Lemma 5.11(b) with $\mathcal{F} = \mathcal{B}$. From (5.46) it is clear that a recurrent point is nonwandering. A minimal point is recurrent and so is nonwandering.

(c): By Corollary 5.14 the set of $x \in X$ such that $S^*x = \mathcal{N}^*\Phi(x)$ is a dense G_δ. Since $x \in \mathcal{N}^*\Phi(x)$ for all x, it follows that $RECUR$ contains this dense G_δ. We leave to the reader the check that with F varying over a countable generating set for $k\mathcal{B}$ and W varying over a countable base of open neighborhood of the diagonal $1_X \subset X \times X$

$$(5.48) \qquad RECUR = \bigcap_{F,W} \bigcup_{g \in F} (\Delta\Phi^g)^{-1}(W),$$

where $\Delta\Phi^g : X \to X \times X$ is defined by $x \mapsto (x, gx)$. This shows that $RECUR$ itself is a G_δ in any case.

(b): As usual, we write X as a surjective inverse limit of metric systems with projections $\pi_\alpha : X \to X_\alpha$. From the assumption and Lemma 5.11(d) it follows that every point is nonwandering in X_α. By part (c) above, the recurrent points are dense in each X_α. Now if U is opene in X then by definition of the inverse limit there exists $U_\alpha \subset X_\alpha$ opene with $(\pi_\alpha)^{-1}(U_\alpha) \subset U$. There exists a recurrent point $y \in U_\alpha$ and an idempotent $u \in S^*$ such that $uy = y$. Since π_α is surjective there exists x such that $\pi_\alpha(x) = y$ and so $\pi_\alpha(ux) = y$. Since u is an idempotent in S^*, ux is a recurrent point in $(\pi_\alpha)^{-1}(U_\alpha) \subset U$. Thus, the recurrent points are dense. For future reference notice that this argument proves that density of the recurrent points is preserved by surjective inverse limit. $\square$

For a classical action $\Phi : S \times X \to X$ we define the *regional proximality relation*

$$(5.49) \qquad QPROX(\Phi) =_{def} (\mathcal{N}(\Phi \times \Phi))^{-1}(1_X).$$

Proposition 5.16. Let $\Phi : S \times X \to X$ be a classical Ellis action of a classical Ellis semigroup (S, G, S^*).

(a) The following conditions on a point $(x, y) \in X \times X$ are equivalent.
  (i) $(x, y) \in QPROX(\Phi)$.
  (ii) There exists $z \in X$ such that $(z, z) \in \mathcal{N}^*(\Phi \times \Phi))(x, y)$.
  (iii) For all open $U, V \subset X$ and $W \subset X \times X$ such that $(x, y) \in U \times V$ and $1_X \subset W$, $N^{\Phi \times \Phi}(U \times V, W)$ is nonempty.

(b) $QPROX(\Phi)$ is a closed relation which contains $PROX(\Phi)$. In particular, if $PROX(\Phi)$ is dense in $X \times X$ then $QPROX(\Phi) = X \times X$.

(c) If X is metrizable and $QPROX(\Phi) = X \times X$ then $PROX(\Phi)$ is a dense, G_δ subset of $X \times X$.

(d) If X is minimal, G is separable and $QPROX(\Phi) = X \times X$ then $PROX(\Phi)$ is dense in $X \times X$.

PROOF. (a), (i) $\Leftrightarrow$ (ii): By (5.44), $(z,z) \in \mathcal{N}(\Phi \times \Phi))(x,y)$ iff $(z,z) \in \mathcal{N}^*(\Phi \times \Phi))(x,y)$ or $(z,z) \in S(x,y)$. In the latter case, $(pz,pz) \in S^*(x,y) \subset \mathcal{N}^*(\Phi \times \Phi))(x,y)$ for all $p \in S^*$.

(i) $\Rightarrow$ (iii): Obvious.

(iii) $\Rightarrow$ (i): We prove the contrapositive. If $\mathcal{N}(\Phi \times \Phi))(x,y)$ is disjoint from the diagonal then this closed set is disjoint from the closure $\overline{W}$ of some open neighborhood W of the diagonal. By the continuity result (5.35) there exist open U, V with $(x,y) \in U \times V$ such that $\mathcal{N}(\Phi \times \Phi))(U \times V)$ is disjoint from $\overline{W}$. A fortiori, $S(U \times V) \cap W = \emptyset$, i.e. $N(U \times V, W) = \emptyset$.

(b): This is clear from (5.46) and the observation that the image of a closed set under a closed relation is closed.

(c): By Corollary 5.14 the set $\{(x,y) : \mathcal{N}(\Phi \times \Phi)(x,y) = S(x,y)\}$ is a dense G_δ because X is metrizable. If $QPROX = X \times X$ then this set is contained in $PROX$. On the other hand, letting W vary over a countable base for open neighborhoods of the diagonal, we have

(5.50) $$PROX = \bigcap_W \bigcup_{g \in G} (\Phi^g \times \Phi^g)^{-1}(W).$$

Thus, $PROX$ is always a G_δ for a classical action on a metrizable space.

(d): Reduce to the metrizable case as in Proposition 5.15(b). Given opene $U, V \subset X$ there exists an index α and opene $U_\alpha, V_\alpha \subset X_\alpha$ such that $(\pi_\alpha \times \pi_\alpha)^{-1}(U_\alpha \times V_\alpha) \subset U \times V$. By (5.43) $QPROX(\Phi_\alpha) = X_\alpha \times X_\alpha$ and so by part (c), there exists a proximal pair $(x,y) \in U_\alpha \times V_\alpha$. Since X is minimal, X_α is and so by Lemma 2.6 there is a proximal pair in $X \times X$ which maps onto (x,y). By choice of U_α and V_α this pair lies in $U \times V$. □

We turn now to notions of transitivity.

Recall that an Ellis action $\Phi : S \times X \to X$ is called *point transitive* iff $TRANS = \{x : Sx = X\}$ is nonempty. We call it S^* *point transitive* when the restricted S^* action is point transitive, i.e. $TRANS_{S^*} = \{x : S^*x = X\}$ is nonempty.

Proposition 5.17. Let (S, G, S^*) be a classical Ellis semigroup and let $\Phi : S \times X \to X$ be a point transitive classical Ellis action.

(a) A point $x \in X$ is an S^* transitive point iff it is a recurrent transitive point. If such a point exists then Φ is S^* point transitive with

(5.51) $\qquad TRANS = TRANS_{S^*}.$

Otherwise, Φ is not S^* point transitive, i.e. $TRANS_{S^*} = \emptyset$.

(b) If X is minimal then Φ is S^* point transitive with $TRANS_{S^*} = X$.

PROOF. (a): Lemma 4.1(b) with $L = S^*$ implies that $TRANS \cap RECUR = TRANS_{S^*}$. Assume that $x \in TRANS_{S^*}$ and $x_1 \in TRANS$. There exists $p \in S$ such that $px_1 = x$. For any $y \in X$ there exists $q \in S^*$ such that $y = qx = qpx_1$. However, (5.3) implies that $qp \in S^*$ and so x_1 is an S^* transitive point.

(b): Since S^* is a closed ideal, minimality of X implies $S^*x = X$ for all $x \in X$. $\square$

For classical actions there is a more general notion of transitivity.

We call the classical Ellis action $\Phi : S \times X \to X$ *transitive* when for every pair of opene subsets $U, V \subset X$, the hitting time set $N^\Phi(U, V)$ is nonempty or, equivalently, $N^\Phi(U, V) \in \mathcal{P}_+$. We call the system Φ S^* *transitive* when for every pair of opene subsets $U, V \subset X$, $N^\Phi(U, V) \in \mathcal{B}$. In general, if $\mathcal{F}$ is a proper family on G then we call the system Φ $\mathcal{F}$ *transitive* when $N^\Phi(U, V) \in \mathcal{F}$ for all opene U, V. For example, a system is called *mixing* when it is $k\mathcal{B}$ transitive. When $S^* = S \setminus G$ this says that each $N^\Phi(U, V)$ is cobounded.

Theorem 5.18. Let (S, G, S^*) be a classical Ellis semigroup and let $\Phi : S \times X \to X$ be a classical Ellis action.

(a) The following conditions are equivalent.
 (1) The system Φ is transitive.
 (2) The prolongation $\mathcal{N}(\Phi)$ equals $X \times X$.
 (3) If an opene subset of X is minus invariant with respect to G then it is dense.
 (4) A proper, closed invariant subset of X is nowhere dense.
Furthermore, Φ is S^* transitive iff $\mathcal{N}^*(\Phi) = X \times X$.

(b) If $TRANS$ is dense in X then the system Φ is transitive as well as point transitive. If, in addition, $TRANS_{S^*} \neq \emptyset$ then the system Φ is S^* transitive as well as S^* point transitive.

(c) If $TRANS$ is nonempty and G invariant then it is dense in X and the system Φ is transitive as well as point transitive. If, in addition, $TRANS_{S^*} \neq \emptyset$ then the system Φ is S^* transitive as well as S^* point transitive.

(d) If X is minimal then Φ is S^* transitive as well as S^* point transitive.

(e) Assume that X is metrizable. If Φ is transitive then $TRANS$ is a dense, G_δ subset of X. In particular, the system is point transitive.

If, in addition, S^* is a G_δ subset of X and Φ is S^* transitive then it is S^* point transitive.

(f) Assume that $\Psi : S \times Y \to Y$ is an Ellis action and that $\pi : X \to Y$ is a surjective, continuous action map. If Φ is transitive (or S^* transitive) then Ψ is a transitive classical Ellis action (resp. an S^* transitive classical Ellis action).

(g) The system Φ satisfies exactly one of the following conditions.

Case (0) Φ is neither transitive nor point transitive.

Case (1) Φ is S^* transitive and S^* point transitive. If (S, G, S^*) satisfies the Surjection Condition then in this case $TRANS = TRANS_{S^*}$ is a dense, G invariant subset and the system Φ has a surjective G action.

Case (2) Φ is S^* transitive but not point transitive, i.e. $TRANS = \emptyset$.

Case (3) Φ is point transitive but not S^* point transitive. In this case, there is a proper closed invariant set $X^* \subset X$ such that $S^*x \subset X^*$ for all $x \in X$ with equality if $x \in TRANS$. Furthermore, for all $x \in TRANS$ $X = Gx \cup X^*$. In particular, $TRANS \cap X^* = \emptyset$.

Case (4) Φ is S^* point transitive but not S^* transitive. In this case, $TRANS = TRANS_{S^*}$ is not a dense subset of X. If (S, G, S^*) satisfies the Surjection Condition then this case does not occur.

(h) Assume that G is separable and that S^* is a G_δ subset of S. If Φ is S^* transitive then the set $RECUR$ of recurrent points is dense in X. If, in addition, (S, G, S^*) satisfies the Surjection Condition then the G action of Φ is surjective.

PROOF. (a), (1)$\Leftrightarrow$ (2): Both conditions say that the hitting time sets $N(U, V)$ are nonempty for all opene U, V. Similarly, all of these sets lie in $\mathcal{B}$ iff Φ is S^* transitive and iff $\mathcal{N}^*(\Phi) = X \times X$.

(1)$\Rightarrow$ (3): Assume that U, V are opene and V is minus invariant. If the system is transitive then U meets $(\Phi^g)^{-1}(V)$ for some $g \in G$. Since V is minus invariant $(\Phi^g)^{-1}(V) \subset V$. Thus, U meets V. As U was arbitrary, V is dense.

(3) $\Leftrightarrow$ (4): An open set is dense iff its complement is nowhere dense. By Proposition 5.6(e) an open set is minus invariant iff its complement is invariant.

(3) $\Rightarrow$ (1): For $V \subset X$ and $A \subset G$ let

$$(5.52) \qquad A^{-1}V \quad =_{def} \quad \bigcup_{g \in A} (\Phi^g)^{-1}(V).$$

If V is opene then $A^{-1}V$ is opene. $G^{-1}V$ is minus invariant. If U is opene then assumption (3) then implies that U meets $G^{-1}V$ and so $N^\Phi(U, V) \neq \emptyset$.

(b): If $TRANS$ is dense then of course the system is point transitive. In addition, if U, V are opene subsets then there exists $x \in U \cap TRANS$. Since

$Sx = X$, G is dense in S and Φ_x is continuous, it follows that Gx is dense in X. Hence, there exist $g \in G$ such that $gx \in V$. Thus, $g \in N^\Phi(U,V)$. If, in addition, the system is S^* point transitive then by Proposition 5.17(a), $TRANS = TRANS_{S^*}$. So we can choose $p \in S^*$ such that $px \in V$. Thus, $(\Phi_x)^{-1}(V)$ is an open subset of S which meets S^* and whose intersection with G is contained in $N^\Phi(U,V)$. By Lemma 5.1(a), the closure of $N^\Phi(U,V)$ contains p and so $N^\Phi(U,V) \in \mathcal{B}$.

(c): If $x \in TRANS$ then Gx is dense in X. So $Gx \subset TRANS$ implies $TRANS$ is dense. The results now follow from (b).

(d): If X is minimal, then $TRANS_{S^*} = X$. Apply (b).

(e): If X is metrizable and Φ is transitive then $TRANS = \{x : Sx = \mathcal{N}\Phi(x)\}$ which is a dense G_δ by Corollary 5.14. Similarly, when Φ is S^* transitive, $TRANS_{S^*} = \{x : S^*x = \mathcal{N}^*\Phi(x)\}$ which is a dense, G_δ when S^* is a G_δ.

(f): Ψ is a classical Ellis action by Proposition 5.4(d). If U, V are opene subsets of Y then $\pi^{-1}(U), \pi^{-1}(V)$ are opene subsets of X because π is continuous and surjective. Furthermore,

$$(5.53) \qquad N^\Psi(U,V) = N^\Phi(\pi^{-1}(U), \pi^{-1}(V)).$$

So Ψ is transitive or S^* transitive when Φ is.

(g): Excluding Case (0) we can assume that Φ is either transitive or point transitive. First, assume the system is S^* point transitive. By Proposition 5.17(a) $TRANS = TRANS_{S^*}$. If the Surjection Condition holds then $g \in G$ implies $S^* = S^*g$. So if $x \in TRANS$ then $X = S^*x = S^*gx$ which implies $gx \in TRANS$. Similarly, $X = S^*x = gS^*x$ implies that Φ^g is surjective. That is, $TRANS$ is G invariant and Φ has a surjective G action. By (c) $TRANS$ is dense and Φ is S^* transitive. This is Case (1). More generally, if $TRANS$ is dense then by (b) Φ is S^* transitive and we are in Case (1).

Thus, if the system is not S^* transitive and so we are in Case (4) instead of Case (1), $TRANS$ cannot be dense the Surjection Condition does not hold.

Assume now that Φ is not S^* point transitive.

If the system is point transitive but not S^* point transitive, i.e. Case (3), then we choose any $x \in TRANS$ and let $X^* = S^*x$. Since $x \notin TRANS_{S^*}$, X^* is a proper closed subset of X. It is invariant because S^* is an ideal. Furthermore, $X = Sx = (G \cup S^*)x = Gx \cup X^*$. If $y \in X$ then $y = px$ for some $p \in S$ implies $S^*y = S^*px \subset S^*x = X^*$ because S^* is a two-sided ideal. Reversing the roles of x and y we see that $X^* = S^*x = S^*y$ for all $y \in TRANS$. Any proper, closed invariant set is disjoint from $TRANS$.

Finally, assume the system is not point transitive, but is transitive. We show that it is S^* transitive and so is in Case (2).

Assume, instead, that Φ is transitive but not S^* transitive. For all x we have $X = \mathcal{N}\Phi(x) = Sx \cup \mathcal{N}^*\Phi(x)$ by part (a) and (5.45). Because Φ is not S^* transitive, there exists $y \in X$ such that $\mathcal{N}^*\Phi(y)$ is a proper, closed subset of X. By Proposition 5.12 this is a proper closed invariant subset of X and so by part (a) transitivity implies it is nowhere dense. Hence, the closed set Sy equals X and the system is point transitive.

(h): If the system is S^* transitive then every point is nonwandering, i.e. $\mathcal{N}^* = X \times X$ implies $|\mathcal{N}^*| = X$. Hence, G separable and S^* a G_δ imply $RECUR$ is dense by Proposition 5.15(b).

When the Surjection Condition holds we return to the proof of that proposition, writing the system as the surjective inverse limit of metrizable systems. Since S^* is assumed a G_δ, parts (f) and (e) imply that these factors are S^* transitive and S^* point transitive, i.e. the factors are in Case (1) of part (g) which implies that the G actions are surjective on the factors. By a standard compactness argument the inverse limit of a family of surjective maps is surjective and so the property of having a surjective G action is also preserved by surjective inverse limits. □

REMARK 5.7. We will see later that there exists a transitive action of a separable classical Ellis semigroup on a nonseparable space. Since Gx is separable it cannot be dense. So Case (2) actions occur because the space is too big. In applications we use Theorem 5.5 and Theorem 5.18(e) to get back to point transitive systems.

Cases (3) and (4) are the odd cases. As noted above Case (4) does not occur if the Surjection Condition Holds. On the other hand, Case (3) does not occur if $S^* = S$. If we replace (S, G, S^*) by (S, G, S) then the transitive systems in Case (3) are thrown into Case (1) and the rest go into Case (4).

Case (3) can occur with the system transitive or not as is seen by the translation actions on $\beta\mathbb{Z}$ and on $\beta\mathbb{Z}_+$, respectively. With an additional assumption on the semigroup we can exclude one peculiarity at least, avoiding S^* transitive actions in Case (3).

Proposition 5.19. Assume that for a classical Ellis semigroup (S, G, S^*) the restriction of the multiplication map M to $S \times G$ is continuous. If $\Phi : S \times X \to X$ is a classical Ellis action which is point transitive and S^* transitive then it is S^* point transitive.

PROOF. We assume that Φ is point transitive but not S^* point transitive and we prove that it is not S^* transitive. Let $x \in TRANS$. Since $x \notin TRANS_{S^*}$, $X = Sx = Gx \cup S^*x$ and S^*x is a closed, proper subset of X. Let U be an opene subset of X with closure $\overline{U}$ disjoint from S^*x. Since $K = (\Phi_x)^{-1}(\overline{U})$ is a subset of G which is closed in S and disjoint from S^*, it is a compact subset of $S \setminus S^* \subset G$.

If $g \in N(U, U)$ then there exists $y \in U$ such that $gy \in U$. We can choose $k \in K$ such that $kx = y$ and so $gkx \in U$. It follows that $gk \in K$. That is,

$N(U,U)$ is contained in

(5.54) $$\hat{K} \quad =_{def} \quad \bigcup_{k \in K} M_k^{-1}(K).$$

It will suffice to show that $\hat{K}$ has closure disjoint from S^*. It then follows that $N(U,U) \subset \hat{K} \notin \mathcal{B}$ and so Φ is not S^* transitive.

Since the restriction of multiplication to $S \times K$ is continuous and since $M(S^* \times K) \subset S^*$, compactness of K implies there exists a neighborhood A of S^* in S such that $M(A \times K) \subset S \setminus K$. Hence, $\hat{K} \subset S \setminus A$. □

Theorem 5.20. Let (S, G, S^*) be a separable, classical Ellis semigroup and let $\Phi : S \times X \to X$ be a transitive classical Ellis action.
  (a) If X is recurrent point minimal then X is minimal. In particular, a transitive, semidistal dynamical system is minimal.
  (b) If the minimal points are dense in X and all $[Min(S)]$ recurrent points are minimal then X is minimal. In particular, a transitive, $[Min(S)]$ semidistal system with dense minimal points is minimal.

PROOF. (a): We reduce to the metrizable case. By Theorem 5.5 the system Φ is the inverse limit of a family of metrizable systems. As metrizable factors of transitive systems they are transitive and point transitive by Theorem 5.18(e). As factors of recurrent point minimal systems they are recurrent point minimal by Theorem 2.9(a). A point transitive, recurrent point minimal system is clearly minimal. As the inverse limit of minimal systems Φ is minimal, see, e.g. the proof of Akin and Glasner (2001) Theorem 2.3. A semidistal system is recurrent point minimal by Theorem 2.7(c).

(b): The reduction to the metrizable case proceeds as in (a). We apply Theorem 2.9(a) and Theorem 2.7(c) to the restricted $[Min(S)]$ action. Notice that Φ is point transitive with minimal points dense exactly when the $[Min(S)]$ action is point transitive. □

Corollary 5.21. Let (S, G, S^*) be a separable, classical Ellis semigroup.
  (a) Let $\Phi : S \times X \to X$ be a classical action and K be a closed invariant subset of X. Assume that the subsystem $\Phi|K$ is transitive. If Φ is semidistal then K is minimal. If Φ is $[Min(S)]$ semidistal and the minimal points are dense in K then K is minimal.
  (b) Let $\Phi : S \times X \to X$ and $\Psi : S \times Y \to Y$ be classical actions. Let $\pi : X \to Y$ be a continuous action map and K be a closed invariant subset of R_π. Assume that Y is minimal and that the subsystem $\Phi^2|K$ is transitive. If π is semidistal then K is minimal. If π is $[Min(S)]$ semidistal and the minimal points are dense in K then K is minimal.

PROOF. (a): X is recurrent point minimal or $[Min(S)]$ recurrent point minimal by Theorem 2.7(c). The subsystem K inherits this property and we can apply Theorem 5.20 to the subsystem $\Phi|K$.

(b): R_π is recurrent point minimal or $[Min(S)]$ recurrent point minimal by Corollary 2.11(b). Apply Theorem 5.20 to the subsystem $\Phi^2|K$. □

Corollary 5.22. Let (S, G, S^*) be a separable, classical Ellis semigroup, let $\Phi : S \times X \to X$ and $\Psi : S \times Y \to Y$ be classical actions and let $\pi : (X, f) \to (Y, g)$ be a surjective, continuous action map.
  (a) If Φ is transitive, Y is minimal and π is semidistal then X is minimal.
  (b) If Φ is transitive with dense minimal points, Y is minimal and π is $[Min(S)]$ semidistal then X is minimal.

PROOF. (a): If Y is minimal then it is recurrent point minimal. If π is semidistal then Corollary 2.15 implies that X is recurrent point minimal. Apply Theorem 5.20(a).

(b): As in part (a). Apply Corollary 2.15 to the restricted $[Min(S)]$ action and then apply Theorem 5.20(b). □

A pair of classical actions $\Phi : S \times X \to X$ and $\Psi : S \times Y \to Y$ is called *weakly disjoint* if the product system $\Phi \times \Psi$ on $X \times Y$ is transitive. This requires that both systems be transitive. A system $\Phi : S \times X \to X$ is called *weakly mixing* if it is weakly disjoint from itself, i.e. the product system $\Phi \times \Phi$ on $X \times X$ is transitive. Φ is called *weak mixing of all orders* if the product systems Φ^n on X^n is transitive for every positive integer n. If X is minimal and Ψ is transitive, then the pair is called *disjoint* if the projection map $\pi_2 : X \times Y \to Y$ is a minimal action map. If both X and Y are minimal then by Proposition 2.13(b) the the pair is disjoint iff the product $X \times Y$ is minimal.

Following Blanchard et al. (2000) and Akin and Glasner (2001) we call a system Φ *scattering* if it is weakly disjoint from every minimal system.

Proposition 5.23. Let (S, G, S^*) be a classical Ellis semigroup and let $\Phi : S \times X \to X$ and $\Phi_1 : S \times X_1 \to X_1$ be classical actions. Assume Φ_1 is transitive and that the minimal points are dense in X_1. If Φ is scattering then Φ_1 and Φ are weakly disjoint.

PROOF. Let U, V and U_1, V_1 be opene subsets of X and X_1, respectively. Because Φ_1 is transitive there exists $g \in G$ such that $W = U_1 \cap (\Phi_1^g)^{-1}(V_1) \neq \emptyset$. Since the minimal points are dense, we can choose a minimal subset $M \subset$

X_1 which meets W and so meets U_1. Since $g_1 M \subset M$, we see that M meets V_1 as well. Since the restriction $\Phi_1|M$ is minimal and Φ is scattering, the product system is transitive on $X \times M$. Hence, $N(U \times (U_1 \cap M), V \times (V_1 \cap M))$ is nonempty. Since the original choice of opene subsets was arbitrary, it follows that the product system on $X \times X_1$ is transitive. $\square$

Let (S, G, S^*) be a classical Ellis semigroup. Following Akin and Glasner (2001) we call a property P of classical (S, G, S^*) actions *residual* if

(i) Any trivial action satisfies P.
(ii) Any factor of a property P system satisfies P.
(iii) Any lift by an irreducible action map of a property P system satisfies P.
(iv) Any inverse limit of a surjective family of property P systems satisfies P.

We will use the symbolism $\Phi \in P$ to mean the system Φ satisfies P. For properties P and P_1 we call P_1 *stronger than* P if P_1 implies P. Given properties P and P_1, and classical Ellis actions $\Phi : S \times X \to X$ and $\Phi_1 : S \times X_1 \to X_1$, we define properties, P', P_{Φ_1}, and P_{P_1} by:

(5.55)
$$\begin{aligned} \Phi \in P' &\iff \Phi^2 : S \times X \times X \to X \times X \in P. \\ \Phi \in P_{\Phi_1} &\iff \Phi \times \Phi_1 : S \times X \times X_1 \to X \times X_1 \in P. \\ \Phi \in P_{P_1} &\iff \Phi \in P_{\Phi_1} \text{ for all } \Phi_1 \in P_1. \end{aligned}$$

Recall that for $\mathcal{F}$ a proper family on G then we call the system Φ $\mathcal{F}$ transitive when $N^\Phi(U, V) \in \mathcal{F}$ for all opene $U, V \subset X$. Thus, transitivity is $\mathcal{P}_+$ transitivity and S^* transitivity is $\mathcal{B}$ transitivity.

Proposition 5.24. (a) The conjunction of any finite or infinite collection of residual properties is a residual property.
(b) If P is a residual property then P' is a residual property stronger than P.
(c) If Φ_1 satisfies the residual property P then P_{Φ_1} is a residual property stronger than P.
(d) If P and P_1 are residual properties with P_1 stronger than P then P_{P_1} is a residual property stronger than P.
(e) Minimality of X is a residual property for $\Phi : S \times X \to X$.
(f) Density of the minimal points in X is a residual property for $\Phi : S \times X \to X$.
(g) Density of the recurrent points in X is a residual property for $\Phi : S \times X \to X$.
(h) $\mathcal{F}$ transitivity is a residual property for any proper family $\mathcal{F}$ on G.
(i) If Φ_1 is transitive then weak disjointness from Φ_1 is a residual property stronger than transitivity.

(j) If Φ_1 is transitive then disjointness from Φ_1 is a residual property stronger than transitivity.

(k) Weak mixing and weak mixing of all orders are residual properties stronger than transitivity.

(l) Scattering (i.e. weak disjointness from all minimal systems) is a residual property stronger than transitivity.

PROOF. The residual properties which were introduced and studied in Akin and Glasner (2001) were all for classes of surjective maps. That is, the results were proved for the special case of Example (4) $(S, G, S^*) = (\beta \mathbb{Z}_+, \mathbb{Z}_+, \beta^* \mathbb{Z}_+)$ with surjective actions. (For this semigroup transitive actions are always surjective as we will see below). However, the proofs easily extend to this general case. We will merely indicate which parts of the above statement correspond to which theorems of Akin and Glasner (2001).

Parts (a)-(d) are proved in Theorem 1.3, (e)-(h) are in Theorem 2.3. For (g) use (2.18) to construct a proof analogous to the one for (f). (i),(j) are in Theorem 2.6 and (l) is in Theorem 2.7. Condition (k) follows from (b). We leave to the reader the easy adjustments or the alternative suspension of disbelief. □

Proposition 5.25. Assume that (S, G, S^*) satisfies the Surjection Condition. Let $\Phi : S \times X \to X$ be a classical action. Each of the following conditions on Φ is a residual property.

(a) Φ is minimal and is a proximal extension of a distal system.

(b) Φ is minimal and is an irreducible extension of a distal system.

(c) Φ is minimal and is a proximal extension of an equicontinuous system.

(d) Φ is minimal and is an irreducible extension of an equicontinuous system.

PROOF. By Corollary 5.10 and the Remark thereafter, the Surjection Condition implies that Φ is a proximal extension of a distal system iff $PROX(\Phi)$ is closed. Now the proof of Akin and Glasner (2001) Theorem 3.6 extends to classical (S, G, S^*) actions. □

We apply Proposition 5.24 to extend Theorem 3.7 of Akin and Glasner (2001) to the semidistal case.

Theorem 5.26. Let (S, G, S^*) be a separable classical Ellis semigroup and let $\Phi : S \times X \to X$ and $\Psi : S \times Y \to Y$ be classical actions. Assume that Ψ is a minimal system and is an irreducible lift of a semidistal system. If Φ

is weakly disjoint from Ψ then it is disjoint from Ψ. That is, the projection map $\pi_1 : X \times Y \to X$ is a minimal map.

PROOF. We begin with the argument from Akin and Glasner (2001) which uses residuality of disjointness and weak disjointness to reduce to the metric case. Notice that since Φ is weakly disjoint from Ψ it is transitive.

If the theorem fails then there is a Φ which is transitive and weakly disjoint from Ψ but not disjoint from Ψ. By Theorem 5.5 Φ is a surjective limit of metric systems and, as factors of Φ, all of these systems are transitive and weakly disjoint from Ψ. If they were all disjoint from Ψ then the inverse limit Φ would be as well. Thus, we can assume that our counterexample, Φ is metric. Furthermore, Φ is weakly disjoint from and is not disjoint from any irreducible factor of Ψ. Thus, we can assume that Ψ is itself is semidistal. Apply Theorem 5.5 to get Ψ as an inverse limit of metric systems and so as before we can assume that Ψ too is metric. The metric factors of the minimal system are minimal because minimality is a residual property. The factors are semidistal by Theorem 2.9(b).

Thus, we need only prove the theorem under the additional assumptions that X and Y are metric and that Ψ is semidistal. Weak disjointness says that the product system is transitive on $X \times Y$. Because $X \times Y$ is metric, Theorem 5.18(e) implies that the product system is point transitive. Since Ψ is a semidistal system the projection map π_1 is a semidistal map. By Theorem 2.14 the map π_1 is a minimal map. □

Corollary 5.27. Assume that (S, G, S^*) is a separable classical Ellis semigroup. For a classical action $\Phi : S \times X \to X$ let P be the property that X is minimal and Φ is disjoint from all scattering systems. P is a residual property and every minimal, semidistal system satisfies P. If $\Psi : S \times Y \to Y$ is a classical action with Y minimal and metric and there exists a continuous, semidistal action map $\pi : Y \to X$ with $\Phi : S \times X \to X$ satisfying P, then Ψ satisfies P.

PROOF. P is a residual property by Proposition 5.24. By Theorem 5.26 all minimal, semidistal systems satisfy P.

Now assume that Ψ is a metric, minimal system which is a semidistal lift of a system Φ in P as in the statement. As usual it suffices to show that if $\Theta : S \times Z \to Z$ is a scattering system with Z metric, then Ψ is disjoint from Θ. That is, the action map $\pi_1 : Z \times Y \to Z$ is a minimal map. Observe that $1_Z \times \pi : Z \times Y \to Z \times X$ is a semidistal map since π is. Since X is minimal and Z is scattering, the product action is transitive on $Z \times Y$. Since $Z \times Y$ is metrizable, the product action is point transitive by Proposition 5.18(e). By Theorem 2.14 the map $1_Z \times \pi$ is minimal. On the other hand,

$\pi_1 : Z \times X \to Z$ is a minimal map because Φ satisfies P. So the composition $\pi_1 : Z \times Y \to Z$ is a minimal map by Proposition 2.13(a). $\square$

In the following chapters we consider the special results which are obtained when we assume that G is a group or an abelian semigroup. We conclude with some odd results which will be useful in each of these sections.

Definition 5.28. Let $\Phi : S \times X \to X$ be a classical action for (S, G, S^*) is a classical Ellis semigroup. We call the action Φ *bizarre* if there is a proper, closed invariant $X^* \subset X$ such that

(5.56) $\qquad\qquad S^* x \quad \subset \quad X^* \qquad \text{for all} \quad x \in X.$

For every point $x \in X \setminus X^*$ there exists a neighborhood U of x and a compact $K \subset S \setminus S^*$ such that

(5.57) $\qquad\qquad Ky \quad = \quad X \qquad \text{for all} \quad y \in U.$

Clearly, a bizarre example is point transitive but not S^* point transitive. The definition is a peculiar sharpening of the general situation described in Case (3) of Theorem 5.18 (g).

Lemma 5.29. *Let (S, G, S^*) be a classical semigroup. Let $\Phi : S \times X \to X$ be a classical action which is neither S^* transitive nor S^* point transitive. Either the action Φ is bizarre or the action Φ is not weak mixing.*

PROOF. Because the system is not S^* transitive there exist opene $U_1, V_1 \subset X$ and compact $K \subset S \setminus S^* \subset G$ such that

(5.58) $\qquad\qquad N(U_1, V_1) \quad \subset \quad K.$

Now there are two possibilities:

Case 1: For all $x \in U_1$, $Kx = X$. Let O be defined as the set of points $z \in X$ such that there exists a neighborhood U of z and a compact $L \subset S \setminus S^*$ such that $Ly = X$ for all $y \in U$. By the case (1) assumption on U_1, the open set $O \subset X$ is nonempty. Clearly, if $z = g\tilde{z} \in O$ for some $\tilde{z} \in X$ and $g \in G$, we can let $\tilde{U} = (\Phi^g)^{-1}(U)$ and $\tilde{L} = Lg$ to see that $\tilde{z} \in O$. That is, O is G minus invariant and so its complement, which we will call X^* is closed and G invariant and so is invariant. If for any $p \in S^*$ and $x \in X$, we had $z = px \in O$, then $Lpx = X$ and since $Lp \subset S^*$, x would be an S^* transitive point. By hypothesis, such points do not exist and so $S^* x \subset X^*$ for all $x \in X$. By Definition 5.28, the action Φ is bizarre.

Case 2: There exists $x \in U_1$ such that Kx is a proper closed subset of X. Let $V_2 \subset X$ be an opene set with closure disjoint from Kx. By continuity of the G action and compactness of K, there exists a neighborhood U_2 of x such that $KU_2 \cap \overline{V_2} = \emptyset$, and so

(5.59) $\qquad N(U_2, V_2) \cap K \;=\; \emptyset.$

From (5.58) and (5.59) it follows that

(5.60) $\quad N(U_1 \times U_2, V_1 \times V_2) \;=\; N(U_1, V_1) \cap N(U_2, V_2) \;=\; \emptyset.$

Thus, the product action $\Phi^2 : S \times X \times X \to X$ is not transitive, i.e. Φ is not weak mixing. $\qquad\square$

Bizarre actions do exist. For example, let $S = \mathbb{Z}_+ \cup \{\infty\}$ be the one point compactification of $\mathbb{Z}_+$ described in Example (4). Let $X = \{x_1, x^*\}$ and let S act on X by $px = x^*$ for all $x \in X$ and all $p \neq 0$ in $\mathbb{Z}_+ \cup \{\infty\}$. Let 0 act as the identity.

Example (5) Let X consist of three points, $X = \{x_1, x_2, x^*\}$. Let S_0 consist of the semigroup of maps $f : X \to X$ such that $f(x^*) = x^*$ and the number of points in $f(X)$ is at most two. This is a two-sided ideal in the semigroup of maps on X which fix x^* which acts on X by evaluation. Let $\mathbb{Z}_+ \cup \{\infty\}$ be the one point compactification of $\mathbb{Z}_+$. Let $\mathbb{Z}_+ \cup \{\infty\}$ act on X by $px = x^*$ for all $x \in X$ and all $p \neq 0$ in $\mathbb{Z}_+ \cup \{\infty\}$. Let 0 act as the identity. We define $S = S_0 \cup \mathbb{Z}_+ \cup \{\infty\}$ with $f \cdot p = p \cdot f = p$ for all $p \neq 0$ in $\mathbb{Z}_+ \cup \{\infty\}$. Let 0 be the identity on S. Let $G = S_0 \cup \mathbb{Z}_+$ and $S^* = \{\infty\}$. (S, G, S^*) is a classical Ellis semigroup and we have defined a classical action Φ on X which is bizarre but now Φ^2 is point transitive with $TRANS = \{(x_1, x_2), (x_2, x_1)\}$. However, Φ is not transitive.

CHAPTER 6

Classical Actions: The Group Case

Throughout this section (S, G, S^*) will be a classical Ellis semigroup with G a group. Since a group acts by homeomorphisms, (S, G, S^*) satisfies the Surjection Condition and for any classical action $\Phi : S \times X \to X$ the G action is surjective. Furthermore, $A \subset X$ is G invariant iff it is G minus invariant iff $gA = A$ for all $g \in G$. Recall that G is assumed to be locally compact (see Definition 5.2).

Proposition 6.1. Let (S, G, S^*) be a classical Ellis semigroup with G a group.

(a) G is a topological group, i.e. the inversion map on G is continuous.
(b) If S^* is a proper subset of S then $S^* = S \setminus G$. Otherwise, $S^* = S$.
(c) If G is a compact group then $G = S = S^*$.
(d) If G is separable, then it is σ-compact and S^* is a G_δ subset of S.
(e) G is metrizable iff it is first countable. If G is metrizable, then it is separable iff it is σ-compact.
(f) If S^* is a proper subset of S then it is a G_δ subset of S iff G is σ-compact.
(g) The identity map on G extends to a unique continuous map $\gamma : \beta_u G \to S$ and γ is a surjective semigroup homomorphism. If S^* is a proper subset of S then

(6.1) $$\gamma^{-1}(S^*) = \beta_u^* G.$$

PROOF. (a): That inversion is continuous is a theorem of Ellis. See Ellis (1957_1) or (1957_2) Theorem 2.

(b): If $g_0 \in S^* \cap G$ and $g \in G$ then $g = (gg_0^{-1})g_0$ lies in the ideal S^*. Thus, if S^* meets G, it contains G. The result then follows because $S = G \cup S^*$.

(c): If G is compact then $G = S$ since G is dense. By (b), $S^* = S$.

(d): Choose a symmetric open, neighborhood U of the identity with compact closure. The union of the sets of n-fold products, $G_0 = \bigcup_n U^n$, is an open subgroup of G which is σ-compact because U is bounded and multiplication is continuous. The left cosets of G_0 form an open partition of G. Since G is separable there are only countably many cosets. As each is σ-compact the union is as well. It follows that $S \setminus G$ is G_δ. Since S is a G_δ subset of itself, (b) implies that S^* is always a G_δ.

(e): A metrizable space is always first countable. On the other hand, from a countable base for the neighborhood system for the identity e we obtain, via (5.7), a countable base for the right-invariant uniformity $\mathcal{U}_G$ described in Example (1) of Chapter 5. By Kelley (1955) Theorem 6.13 the uniformity is metrizable. In fact, the construction given in the proof there yields a right-invariant metric.

A metrizable σ-compact space is always separable. On the other hand, for G separability implies σ-compactness by (d).

(f): By (b), S^* is the complement of G in S. If G is σ-compact then it is an F_σ subset of S and so has a G_δ complement. On the other hand, if $S \setminus G$ is the intersection of a sequence of open sets $\{U_n\}$ then each U_n has a complement which is contained in G and is closed in S and so is compact. Hence, as the union of the complements of the U_n's the set G is σ-compact.

(g): As described in Example (1) of Chapter 5, the topological action of G on S by translation extends to $\Psi : \beta_u G \times S \to S$, a classical Ellis action of $(\beta_u G, G, \beta_u^* G)$. Let $\gamma = \Psi_e$. This is a continuous map which extends the identity. It is unique and surjective because G is dense in both semigroups. Furthermore, the equation $\gamma(p_1 p_2) = \gamma(p_1)\gamma(p_2)$ holds when $p_1, p_2 \in G$. Hence, for each $p_1 \in G$ continuity and density imply that the equation holds for all $p_2 \in \beta_u G$. Finally, with p_2 fixed in $\beta_u G$ continuity and density imply that it holds for all $p_1 \in \beta_u G$ as well.

Since γ is surjective, $\gamma^{-1}(S \setminus G) \subset \beta_u^* G$. On the other hand, if U is an open subset of G which is bounded in G (i.e. it has compact closure in G) then $\gamma^{-1}(U)$ is an open subset of $\beta_u G$ with $G \cap \gamma^{-1}(U) = U$ bounded in G. By Lemma 5.1(b), $G \supset \gamma^{-1}(U)$. It follows that no point of $\beta_u G \setminus G$ maps into G. Finally, $\beta_u^* G = \beta_u G \setminus G$ unless G is compact. When G is compact, γ is just the identity map on G. $\qquad\square$

REMARK 6.1. If $\Phi : S \times X \to X$ is a classical action of (S, G, S^*) where X is metrizable and G is a group which is not σ-compact then $Sx = S^*x$ for all $x \in X$. This is obvious if $S = S^*$. Otherwise, S^* is not a G_δ by (f) above. As a closed subset of a metrizable space S^*x is a G_δ and so is its inverse image $(\Phi_x)^{-1}(S^*x)$. As it contains S^* as a proper subset $(\Phi_x)^{-1}(S^*x)$ meets G. Hence, Gx meets, and so is contained in, the closed invariant set S^*x. So $Sx = Gx \cup S^*x = S^*x$.

Proposition 6.2. Let (S, G, S^*) be a classical Ellis semigroup with G a group and let $\Phi : S \times X \to X$ be a classical Ellis action.

(a) The prolongations $\mathcal{N}\Phi$ and $\mathcal{N}^*\Phi$ are closed, invariant subsets of $X \times X$. In fact, for all $p, q \in S$

(6.2) $\qquad (\Phi^p \times \Phi^q)(\mathcal{N}\Phi) \subset \mathcal{N}\Phi \qquad$ and $\qquad (\Phi^p \times \Phi^q)(\mathcal{N}^*\Phi) \subset \mathcal{N}^*\Phi$

6. CLASSICAL ACTIONS: THE GROUP CASE

with equality if $p, q \in G$. Furthermore, if either p or q is in S^* then
(6.3) $$(\Phi^p \times \Phi^q)(\mathcal{N}\Phi) \subset \mathcal{N}^*\Phi.$$

(b) For all $x \in X$ and $g \in G$
(6.4) $$\begin{aligned} \mathcal{N}\Phi(gx) &= \mathcal{N}\Phi(x) = g\mathcal{N}\Phi(x). \\ \mathcal{N}^*\Phi(gx) &= \mathcal{N}^*\Phi(x) = g\mathcal{N}^*\Phi(x). \end{aligned}$$

(c) Both prolongation relations are symmetric. That is,
(6.5) $$\mathcal{N}\Phi = (\mathcal{N}\Phi)^{-1} \quad \text{and} \quad \mathcal{N}^*\Phi = (\mathcal{N}^*\Phi)^{-1}.$$

(d) The nonwandering set $|\mathcal{N}^*\Phi|$ is a closed, invariant subset of X with
(6.6) $$S^*x \subset |\mathcal{N}^*\Phi| \quad \text{for all} \quad x \in X.$$
If x is a nonwandering point then $\mathcal{N}\Phi(x) = \mathcal{N}^*\Phi(x)$.

(e) If x is a wandering point, i.e. $x \in X \setminus |\mathcal{N}^*\Phi|$ then Iso_x is a compact subgroup of G and if K is any neighborhood of Iso_x in G, there exists an open $U \subset X$ with $x \in U$ such that
(6.7) $$g_1 U \cap g_2 U \neq \emptyset \quad \Longrightarrow \quad g_2^{-1} g_1 \in K.$$

(f) The regional proximity relation $QPROX(\Phi)$ is a closed, invariant subset of $X \times X$.

PROOF. Notice that if $S^* = S$ then $\mathcal{B} = \mathcal{P}_+$ and $\mathcal{N}^*\Phi = \mathcal{N}\Phi$ for any classical action Φ. Otherwise, $S^* = S \setminus G$ by Proposition 6.1(b). So in that case, $\mathcal{B}$ consists exactly of the unbounded subsets of G.

(a), (b), (c): Observe that for $U, V \subset X$ and for $g_1, g_2 \in G$
(6.8) $$\begin{aligned} N^\Phi(V, U) &= N^\Phi(U, V)^{-1}, \\ N^\Phi(g_1 U, g_2 V) &= g_2 N(U, V) g_1^{-1}. \end{aligned}$$

Furthermore,
(6.9) $$F \in \mathcal{B} \quad \Longrightarrow \quad F^{-1}, \; g_2 F g_1^{-1} \in \mathcal{B}.$$

Symmetry in (6.5) follows as do the inclusions of (6.2) when $p, q \in G$. Then (6.2) follows for all $p, q \in S$ because G is dense in S, the maps Φ_x are continuous and the prolongations are closed. Equality when $p, q \in G$ is obtained by applying the result with p^{-1}, q^{-1}. The equations of (6.4) are special cases with applied to g and the identity e, e.g. $(x, y) \in \mathcal{N}^*$ iff $(x, gy) \in \mathcal{N}^*$. Finally, suppose that $(x, y) \in \mathcal{N}$ and $q \in S^*$. By (5.44), $\mathcal{N}(x) = Sx \cup \mathcal{N}^*(x)$. Hence, $qy \in qSx \cup q\mathcal{N}^*(x)$. Because $qSx \subset S^*x \subset \mathcal{N}^*(x)$ and $q\mathcal{N}^* \subset \mathcal{N}^*$ by (6.2), we have $(x, qy) \in \mathcal{N}^*$. By (6.2) again $(px, qy) \in \mathcal{N}^*$. The case when $p \in S^*$ instead follows from this case and the symmetry equations (6.5).

(d): Since $S^*x \subset \mathcal{N}^*(x)$, (6.2) implies that $(px, y) \in \mathcal{N}^*$ for all $p \in S$ and $y \in S^*x$. By symmetry, we have, for any x
(6.10) $$(Sx \times S^*x) \cup (S^*x \times Sx) \subset \mathcal{N}^*\Phi.$$

Intersecting with the diagonal we obtain (6.6). Since $\mathcal{N}^*$ and the diagonal 1_X are closed invariant sets, so is their intersection. Finally, x is a nonwandering point exactly when $x \in \mathcal{N}^*(x)$. As this set is invariant, this implies $Sx \subset \mathcal{N}^*(x)$. So $\mathcal{N}(x) = \mathcal{N}^*(x)$ by (5.44).

(e): If x is wandering then there is an open $U_0 \subset X$ and a compact $K_0 \subset G$ such that $x \in U_0$ and $N(U_0, U_0) \subset K_0$. Notice that $S^* = S$ implies that every point is recurrent and so is nonwandering. So $S^* = S \setminus G$. Then $N(x, U_0) = (\Phi_x)^{-1}(U_0) \cap G$. By Lemma 5.1 this subset of $N(U_0, U_0)$ is dense in the open set $(\Phi_x)^{-1}(U_0)$ and so we have:

$$(6.11) \qquad Iso_x \subset (\Phi_x)^{-1}(U_0) \subset K_0 \subset G.$$

Notice that for any $y \in X$ $g \in G \cap Iso_y$ implies $g^{-1} \in G \cap Iso_y$ and so $g \in G \cap Iso_y$ is a subgroup of G. For our wandering point x, (6.11) implies that Iso_x is a compact subgroup of G. Given a neighborhood K of Iso_x we shrink to assume K is compact and $K \subset K_0$. The G action map $\Phi : G \times X \to X$ is continuous. As A varies over the compact neighborhoods of x which are contained in U_0, the collection of compact sets $\{(\Phi)^{-1}(A) \cap (K_0 \times A)\}$ is closed under finite intersection and has intersection $Iso_x \times \{x\}$. Since $A \subset U_0$, and so $N(A, A) \subset K_0$, $(\Phi)^{-1}(A) \cap (K_0 \times A) = (\Phi)^{-1}(A) \cap (G \times A)$. It follows that we can choose an open set U which contains x and such that

$$(6.12) \qquad (\Phi)^{-1}(U) \cap (G \times U) \subset K \times U_0 \qquad \text{and so} \qquad N^\Phi(U, U) \subset K.$$

It then follows from (6.8) that $g_1 U \cap g_2 U \neq \emptyset$, i.e. $e \in N(g_1 U, g_2 U)$ iff $g_2^{-1} g_1 \in N(U, U)$ which implies $g_2^{-1} g_1 \in K$.

(f): If $(x, y) \in QPROX$ then for some $z \in X$, $(z, z) \in \mathcal{N}^*(\Phi \times \Phi)(x, y)$. By part (a), $(z, z) \in \mathcal{N}^*(\Phi \times \Phi)(px, py)$ for all $p \in S$. Hence, $QPROX$ is invariant. $\square$

REMARK 6.2. (a) The product $(S \times S, G \times G, (S^* \times S) \cup (S \times S^*))$ is a classical Ellis semigroup with $G \times G$ a group. We clearly obtain an action of this product semigroup on $X \times X$. We will say that a subset of $X \times X$ is $S \times S$ *invariant* when it is invariant with respect to this product action. Part (a) says that the prolongation relations are $S \times S$ invariant. However, in part (f) the regional proximality relation is just S invariant.

(b) When G is a separable group, we can obtain the center for a classical action by using a (possibly transfinite) inductive procedure. Since the nonwandering set is invariant we obtain a subsystem by restricting the action to it. The nonwandering set of the subsystem can be a proper subset. Continue restricting inductively, intersecting at the limit ordinals. At each stage the recurrent points of the original system are retained. The process will stabilize at a system with respect to which all of its points are nonwandering. Since G is separable, Proposition 5.15(b) implies that the recurrent points are dense. Thus, we have reached the center $[RECUR]$.

Theorem 6.3. Let (S, G, S^*) be a classical Ellis semigroup with G a group. If $\Phi : S \times X \to X$ is a classical Ellis action, then Φ satisfies exactly one of the following conditions.

Case (0) Φ is neither transitive nor point transitive.

Case (1) Φ is S^* transitive and S^* point transitive. In this case $TRANS = TRANS_{S^*}$ a dense, G invariant subset.

Case (2) Φ is S^* transitive but not point transitive, i.e. $TRANS = \emptyset$. If, in addition, G is separable, then $RECUR$ is a dense, G invariant subset of X. This case does not occur if X is metrizable.

Case (3) Φ is point transitive and transitive but is neither S^* transitive nor S^* point transitive. In this case, $TRANS$ consists of a single G orbit, i.e. $TRANS = Gx$ for $x \in TRANS$, and it is an open, dense, G invariant subset of X. The complement $X^* = X \setminus TRANS$ is a closed, nowhere dense, invariant subset which contains the center $[RECUR]$. Furthermore, $S^*x \subset X^*$ for all $x \in X$ with equality if $x \in TRANS$. The $\mathcal{B}$ prolongation satisfies $\mathcal{N}^*\Phi = (X^* \times X) \cup (X \times X^*)$. Thus, X^* equals the nonwandering set $|\mathcal{N}^*\Phi|$.

PROOF. Recall that for any Ellis action $TRANS$ satisfies the capturing property (see the discussion following Theorem 4.6). This implies that $TRANS$ is G minus invariant. Since G is a group, $TRANS$ is G invariant. By Theorem 5.18(c) any point transitive system is transitive and any S^* point transitive system is S^* transitive.

Next observe that if Φ is point transitive and S^* transitive then it is S^* point transitive. Apply the proof of Proposition 5.19. It suffices to show that $\hat{K}$ defined by (5.54) is a compact subset of G. Since K is compact and G is a topological group this is true because $\hat{K} = K \cdot K^{-1}$.

Because G is a group, the Surjection Condition holds. We apply Theorem 5.18(g), refining our description of the cases. Case (4) does not apply. In Case (2), separability of G implies that S^* is a G_δ by Proposition 6.1(d). Hence, by Theorem 5.18(h) the set of recurrent points is dense.

In Case (3) the system is point transitive and not S^* point transitive. By the above remarks it is transitive but not S^* transitive. Let $x \in TRANS$. From Theorem 5.18(g) Case (3), we have $X = Gx \cup X^*$ and $S^*y \subset X^*$ with equality if $y \in TRANS$. If $g_0 x \in X^*$ for some $g_0 \in G$ then for all $g \in G$, $gx = (gg_0^{-1})g_0 x \in X^*$. Since closed set X^* is a proper subset of X, it follows that $Gx = X \setminus X^*$ and so this set is open. Since $TRANS$ is G invariant all of the points of Gx are transitive. Since X^* is proper, closed and invariant, none of the points of X^* are transitive. Thus, X is the disjoint union of the single orbit $Gx = TRANS$ and the proper, closed invariant set X^* which is nonempty since it contains S^*X. Since Gx is open and dense it follows that X^* is nowhere dense. Clearly, if a closed invariant set meets Gx then it contains it and so it equals X since Gx is dense. Thus, X^* is the maximum among the proper, closed invariant subsets of X.

Because Φ is not S^* transitive in this Case (3) situation, it follows that $\mathcal{N}^*\Phi$ is a proper subset of $X \times X$. If it were to meet any point of $Gx \times Gx$ then by Proposition 6.2(a) it would contain this dense set and so would equal $X \times X$. Hence, we have $\mathcal{N}^*\Phi \subset (X^* \times X) \cup (X \times X^*)$. Since $X^* = S^*y \subset \mathcal{N}^*\Phi(y)$ for all $y \in Gx$, we have $Gx \times X^* \subset \mathcal{N}^*\Phi$. Because the prolongation is closed and Gx is dense this implies $X \times X^* \subset \mathcal{N}^*\Phi$. From symmetry of the relation, i.e. Proposition 6.2(c), we get the equation $\mathcal{N}^*\Phi = (X^* \times X) \cup (X \times X^*)$. Intersecting with the diagonal we see that X^* is the nonwandering set, which always contains the closure of the set of recurrent points. □

When S^* is a proper subset of S then translation action of (S, G, S^*) is a Case (3) action. In Case (1) it can happen that $TRANS_{S^*} = Gx = X$. We call a classical action $\Phi : S \times X \to X$ a *homogeneous space* of (S, G, S^*) when (S, G, S^*) is a classical Ellis semigroup with G a group and Φ is a classical action such that $Gx = X$ for some $x \in X$.

Theorem 6.4. Let (S, G, S^*) be a classical Ellis semigroup with G a group.

(a) If H is a closed subgroup of G, then the space G/H of left cosets of H is a locally compact, Hausdorff space with quotient topology induced by the projection $\pi_H : G \to G/H$ defined by $g \mapsto gH$. The projection π_H is an open map and the action $M_H : G \times G/H \to G/H$ defined by $(g_1, g_2H) \mapsto g_1g_2H$ is continuous and so is a topological action of G on G/H.

G/H is compact iff H is a *syndetic subgroup*, i.e. there exists a compact set $K \subset G$ such that $G = K \cdot H$. If H is a syndetic subgroup and the closure $\overline{H}$ in S is a co-ideal in S then the action of G on G/H extends to a classical action of S on G/H which we will call the translation action on G/H. π_H extends to a continuous map of S onto G/H which is a continuous surjective action map with the translation action of S used on the domain. The action on G/H is minimal.

If H is any syndetic subgroup of G then the action of G on G/H extends to a classical action of $(\beta_u G, G, \beta_u^* G)$ on G/H and π_H extends to a continuous map of $\beta_u G$ onto G/H which is a continuous surjective action map with the translation action of $\beta_u G$ used on the domain.

(b) Assume now that G is σ-compact. Let $\Phi : S \times X \to X$ be a classical Ellis action with $x \in X$ such that $Gx = X$. $H = Iso_x \cap G$ is a syndetic, closed subgroup of G with $\overline{H}$ equal to the co-ideal Iso_x. The map $\Phi_x : S \to X$ factors to define a homeomorphism from G/H onto X which is an action map isomorphism from the translation action on G/H to Φ.

PROOF. Notice first that if A is any subset of G then with $\overline{A}$ the closure in S, $\overline{A} \cap G$ is the closure in G because G has the relative topology. In particular, if H is a closed subgroup of G then

(6.13) $$H = \overline{H} \cap G.$$

Next assume that $\Psi : G \times Y \to Y$ is a topological action with Y locally compact and such that $Gy = Y$ for some y. Let $H = \Psi_y^{-1}(y) = \{h \in G : hy = y\}$. H is clearly, a closed subgroup and $G = K \cdot H$ iff $\Psi_y(K) = Y$. In particular, if H is syndetic, then Y is compact. Conversely, if Y is compact and if there exists a compact subset K_0 of G such that $\Psi_y(K_0)$ has nonempty interior in Y then H is syndetic. For then we can choose a finite collection of translates $g_1 K_0, ..., g_n K_0$ such that $\bigcup_{i=1}^n \Psi_y(g_i K_0) = \bigcup_{i=1}^n g_i \Psi_y(K_0) = Y$. Let $K = \bigcup_{i=1}^n g_i K_0$. If Ψ_y is an open map then we can choose K_0 to be the closure of any bounded, opene subset of G. On the other hand, if G is σ-compact, say $G = \bigcup_{i=1}^\infty K_i$ the compact space Y is the union of the sequence of compact subsets $\Psi_y(K_i)$. By the Baire Category Theorem at least one of the $\Psi_y(K_i)$'s has a nonempty interior. Notice that some assumption is needed because if we replace G by G_{dis} with the discrete topology then even if H is syndetic in G, H_{dis} will not be syndetic in G_{dis} when Y is infinite.

(a): The first part summarizes some standard results about topological groups. If $V \subset G$ is open then $\pi_H^{-1}(\pi_H(V)) = V \cdot H$ which is the union of translates of V and so is open. Hence, $\pi_H(V)$ is open in the quotient topology which says that π_H is an open map. If M is the multiplication on G then $M_H \circ (1_G \times \pi_H) = p_H \circ M$ which is continuous. Since $1_G \times \pi_H$ is an open map it is a quotient map and so M_H is continuous. Clearly, π_H is an action map. If $g \notin H$ then we can choose V an open neighborhood or e such that $V^{-1}Vg \cap H = \emptyset$ because H is closed. It follows that $\pi_H(V)$ and $\pi_H(Vg)$ are disjoint open sets. Hence, G/H is Hausdorff. It is locally compact since π_H is open. Applying the previous paragraph with $y = eH \in Y = G/H$ we see that since $\pi_H = \Psi_y$ is open, G/H is compact iff H is syndetic.

Now assume that K is a compact subset of G such that $K \cdot H = G$ and that $\overline{H}$ is a co-ideal in S. We show that π_H extends to a continuous function on S. For $p \in S \setminus G$ let $g_i = k_i h_i$ be a net in G converging to p. By going to a subnet we can assume that k_i converges to $k \in K$ and h_i converges to $h \in \overline{H}$. Then in G/H the net $g_i H$ converges to kH and we define $\pi_H(p)$ to be this limit point. We show that this extension is well-defined by considering two such nets $g_{i'}^1$ and $g_{i''}^2$ converging to p, with $k_{i'}^1, k_{i''}^2$ converging to points $k^1, k^2 \in K$ and with $h_{i'}^1, h_{i''}^2$ converging to points $h^1, h^2 \in \overline{H}$. Because the action of G on S is continuous

(6.14) $$h^1 = \lim(k_{i'}^1)^{-1} g_{i'} = (k^1)^{-1} p = (k^1)^{-1} \lim g_{i''} = (k^1)^{-1} k^2 h^2.$$

Because $h^1, h^2 \in \overline{H}$ and $\overline{H}$ is a co-ideal, it follows that $(k^1)^{-1}k^2 \in \overline{H} \cap G = H$. Hence, the limit points $k^1 H$ and $k^2 H$ are equal. This argument shows as well that π_H is continuous at each point of $S \setminus G$ and so extends to a continuous map on S (see, e.g. Kelley (1955) Exercise 3D). By Proposition 3.11, the G action on G/H extends to a classical Ellis action of (S, G, S^*) with π_H a surjective action map from the translation action. Since every point is transitive the action is minimal.

In any case, if H is syndetic, G/H is compact and we obtain the extension to $\beta_u G$ as described in Example (1) of Chapter 5.

(b): When G is σ-compact, the subgroup $H = Iso_x \cap G$ is syndetic as described in the second paragraph above. Let $K \subset G$ be compact with $K \cdot H = G$. If $p \in Iso_x$ we can choose $g_i = k_i h_i$ be a net in G converging to p with k_i converging to $k \in K$ and h_i converging to $h \in \overline{H} \subset Iso_x$. This time we know that Φ_x is continuous on S and so $x = px = \lim g_i x = khx$. Since $p, h \in Iso_x$ it follows that $k \in Iso_x \cap G = H$. Since the multiplication M^k is continuous on S it follows that $p = kh \in \overline{H}$. Thus, $Iso_x = \overline{H}$.

The factorization map from G/H to X is defined by $gH \mapsto gx$. Since $g_1 x = g_2 x$ iff $g_1^{-1} g_2 \in H$ and so iff $g_1 H = g_2 H$, the map is a bijection. Since $\pi_H : G \to G/H$ is a quotient map, the factor map is continuous and so is a homeomorphism by compactness. The map and its inverse are clearly action maps. $\square$

REMARK 6.3. The assumption of σ-compactness is necessary in (b). Let X be a compact group, e.g. the circle, and let G be the same group with the discrete topology. We obtain a classical action of $(\beta G, G, \beta^* G)$ on X by translation. For $x = e$, the identity element, $Gx = X$ with $Iso_x \cap G = \{e\}$ but Iso_x is much larger. In Keynes (1967) the author investigates homogeneous systems. He provides algebraic description of the proximal and regionally proximal relations and in particular gives necessary and sufficient conditions for such a system to be distal, see also Ellis (1969) and Auslander (1988). The book *Flows on homogeneous spaces* by L. Auslander, F. Hahn and L. Green (1960) is a pioneering work on the subject and contains one of the first applications of Ellis' theory to the theory of Lie Group actions.

We now deal with a technical difficulty about nonabelian actions. The problem is that for $U, V \subset X$ and $g \in G$ it is usually not true that $N^\Phi(gU, gV) = N^\Phi(U, V)$. Instead, by (6.8)

(6.15) $$N^\Phi(gU, gV) = gN^\Phi(U, V)g^{-1},$$

Let $\Phi : S \times X \to X$ be a classical action of (S, G, S^*). For $U, V \subset X$ and $F \subset G$ let

(6.16) $$N^\Phi_F(U, V) =_{def} \bigcap_{g \in F} N^\Phi(gU, gV) = \bigcap_{g \in F} gN^\Phi(U, V)g^{-1}$$

It will be useful to introduce the Φ *centralizer of* $x \in X$:

(6.17) $\quad Cent(\Phi, x) \quad =_{def} \quad \{p \in S : pgx = gpx \quad \text{for all } g \in G\}.$

Since the action of G is continuous, this is a closed set.

We will call Φ *adjoint transitive* if for every pair of opene subsets U, V of X and for every finite subset F of G, the set $N_F^\Phi(U, V)$ is nonempty, i.e. $N_F^\Phi(U, V) \in \mathcal{P}_+$. We will call Φ S^* *adjoint transitive* if for every pair of opene subsets U, V of X and for every finite subset F of G, the set $N_F^\Phi(U, V) \in \mathcal{B}$.

We say that Φ is $\mathcal{F}$ *adjoint transitive* if $N_F^\Phi(U, V) \in \mathcal{F}$ for every finite $F \subset G$ and opene $U, V \subset X$. Thus S^* adjoint transitivity is the same as $\mathcal{B}$ adjoint transitivity.

For any set A we denote by $Fin(A)$ the set of finite subsets of A.

Proposition 6.5. *Let* (S, G, S^*) *be a classical Ellis semigroup with* G *a group.*

(a) *Any factor of an adjoint transitive system (or an S^* adjoint transitive system) is adjoint transitive (resp. is S^* adjoint transitive).*

(b) *Adjoint transitivity is a residual property stronger than transitivity. S^* adjoint transitivity is a residual property stronger than adjoint transitivity and stronger than S^* transitivity.*

(c) *If G is abelian then a classical action is adjoint transitive (or S^* adjoint transitive) iff it is transitive (resp. iff it is S^* transitive).*

PROOF. (a): The analogue of (5.53) with N replaced by N_F is still true. Use the proof of Theorem 5.18(f).

(b): Using (a), the proof for transitivity from Akin and Glasner (2001) Theorem 2.6 works here, too.

(c): Clearly, (6.16) implies that for all $U, V \subset X$ and $F \subset G$

(6.18) $\quad N_F^\Phi(U, V) \quad = \quad N^\Phi(U, V) \quad$ when G is abelian.

$\square$

The adjoint prolongation relations are defined by analogy with the originals. Let $\mathcal{F}$ be a family of subsets of G.

(6.19) $\quad \mathcal{N}_{\mathcal{F}adj}\Phi \quad = \quad \{(x, y) : N_F^\Phi(U, V) \in \mathcal{F} \quad \text{for all open} \\ U, V \subset X, \quad \text{with} \quad x \in U, y \in V, \quad \text{and all} \quad F \in Fin(G)\}.$

Thus Φ is $\mathcal{F}$ adjoint transitive iff $\mathcal{N}_{\mathcal{F}adj}\Phi = X \times X$.

As before we use the notation:

(6.20) $\quad \mathcal{N}_{adj}\Phi \quad = \quad \mathcal{N}_{\mathcal{P}_+ adj}\Phi \quad$ and $\quad \mathcal{N}_{adj}^*\Phi \quad = \quad \mathcal{N}_{\mathcal{B}adj}\Phi.$

Recall that if $k\mathcal{F}$ is a filter then the hull $H(k\mathcal{F})$ is defined in (5.29) to be the intersection in S of the closures of the elements of $k\mathcal{F}$. As remarked in (5.30) the hull of $k\mathcal{P}_+$ is S and the hull of $k\mathcal{B}$ is S^*.

Lemma 6.6. Let (S, G, S^*) be a classical Ellis semigroup with G a group, $\mathcal{F}$ be a family of subsets of G and let $\Phi : S \times X \to X$ be a classical action.

(a) $\mathcal{N}_{\mathcal{F}adj}\Phi$ is a closed relation on X.

(b) Φ is adjoint transitive (or S^* adjoint transitive) iff $\mathcal{N}_{adj}\Phi = X \times X$ (resp. iff $\mathcal{N}^*_{adj}\Phi = X \times X$).

(c) Assume that $k\mathcal{F}$ is a filter. Let $\mathcal{T}$ be an opene base for the topology of X. For $x, y \in X$

(6.21)
$$y \in (Cent(\Phi, x) \cap H(k\mathcal{F}))x \iff N_F^\Phi(x, V) \in \mathcal{F} \text{ for all } F \in Fin(G), V \in \mathcal{T}_y.$$

where $\mathcal{T}_y$ denotes the set of basic neighborhoods of y. Furthermore, for $F \subset G$ and $V \subset X$

(6.22)
$$N_F^\Phi(x, V) \in \mathcal{F} \iff x \in \bigcup_{g \in A} \bigcap_{h \in F} h^{-1}g^{-1}hV \text{ for all } A \in k\mathcal{F}.$$

(d) With $k\mathcal{F}$ a filter, and $x \in X$:

(6.23)
$$(Cent(\Phi, x) \cap H(k\mathcal{F}))x \subset \mathcal{N}_{\mathcal{F}adj}\Phi(x).$$

On the other hand, $y \notin \mathcal{N}_{\mathcal{F}adj}\Phi(x)$ iff $x \notin \overline{\bigcup_{g \in A} \bigcap_{h \in F} h^{-1}g^{-1}hV}$ for some $F \in Fin(G), A \in k\mathcal{F}$ and $V \in \mathcal{T}$ with $y \in V$.

(e) Assume that $k\mathcal{F}$ is a filter, generated as a family by $\mathcal{A}$. Let $\mathcal{T}$ be an opene base for the topology of X.

(6.24)
$$\{x \in X : \mathcal{N}_{\mathcal{F}adj}\Phi(x) = (Cent(\Phi, x) \cap H(k\mathcal{F}))x\} =$$
$$\bigcap \{X \setminus Bdry[\bigcup_{g \in A} \bigcap_{h \in F} h^{-1}g^{-1}hV] : F \in Fin(G), A \in \mathcal{A}, V \in \mathcal{T}\}.$$

PROOF. (a), (b): Obvious.

(c): If $y = px$ with $p \in Cent(\Phi, x) \cap H(k\mathcal{F})$ then for all $h \in G$, $h^{-1}phx = y$ and so $p \in \bigcap_{h \in F}(\Phi_{hx})^{-1}(hV)$ for every $V \subset X$ with $y \in V$ and every $F \subset G$. If V is open and F is finite, then this set is open and so contains $N_F(x, V)$ as a dense subset by Lemma 5.1(a). The closure in S of every $A \in k\mathcal{F}$ contains p since p is in the hull of $k\mathcal{F}$. Hence, $A \cap \bigcap_{h \in F}(\Phi_{hx})^{-1}(hV) = A \cap N_F(x, V)$ is nonempty for every $A \in k\mathcal{F}$. That is, $N_F(x, V) \in \mathcal{F}$.

The converse is harder. Consider the set $D = Fin(G) \times \mathcal{K}_y \times k\mathcal{F}$ where $\mathcal{K}_y$ is the set of compact neighborhoods of y. This set is directed in the obvious way, increasing the finite sets and decreasing the others. For $(F, K, A) \in D$ the set $N_F(x, K) \in \mathcal{F}$ by assumption and so we can choose $g(F, K, A) \in A \cap N_F(x, K)$. Let p be a limit point in S for this net, indexed by D. For each fixed (F_0, K_0, A_0) we have $g(F, K, A) \in \overline{A_0}$ and $h^{-1}g(F, K, A)hx \in K_0$ for all $h \in F_0$ provided that (F, K, A) follows (F_0, K_0, A_0) in D. Going to the limit, we see that for all (F_0, K_0, A_0) in D, $p \in \overline{A_0}$ and $h^{-1}phx \in K_0$ for all $h \in F_0$. Hence, $p \in H(k\mathcal{F})$ and $h^{-1}phx = y$ for all $h \in G$.

Both conditions in (6.22) say that for every $A \in k\mathcal{F}$ there exists a $g \in A$ such that $h^{-1}ghx \in V$ for all $h \in F$.

(d): We proceed as in the proof of Lemma 5.11(c).

Part (c) implies that for $k\mathcal{F}$ is a filter (6.21) holds which clearly implies (6.23).

Furthermore, part (c) says that $y \in (Cent(\Phi, x) \cap H(k\mathcal{F}))x$ iff $x \in \bigcup_{g \in A} \bigcap_{h \in F} h^{-1}g^{-1}hV$ for all $(F, V, A) \in Fin(G) \times \mathcal{T}_y \times k\mathcal{F}$.

On the other hand, $y \notin \mathcal{N}_{\mathcal{F}adj}\Phi(x)$ iff for some $(U, V) \in \mathcal{T}_x \times \mathcal{T}_y$ and some $F \in Fin(G)$ $N_F(U, V) \notin \mathcal{F}$ and so iff for some such U, V and some $A \in k\mathcal{F}$ $N_F(U, V) \cap A = \emptyset$, or, equivalently, $U \cap (\bigcup_{g \in A} \bigcap_{h \in F} h^{-1}g^{-1}hV) = \emptyset$. Hence, $y \notin \mathcal{N}_{\mathcal{F}adj}\Phi(x)$ iff for some $(F, V, A) \in Fin(G) \times \mathcal{T}_y \times k\mathcal{F}$, $x \notin \bigcup_{g \in A} \bigcap_{h \in F} h^{-1}g^{-1}hV$.

(e): This follows from part (d) just as Lemma 5.11(d) was obtained from Lemma 5.11(c). □

REMARK 6.4. Unfortunately, it does not appear to suffice in (6.21) to allow F to vary only in $Fin(G_0)$ where G_0 is a dense subgroup of G. The problem is that the *adjoint action* $Ad : G \times S \to S$ defined by $Ad(g, p) = gpg^{-1}$ is not usually a continuous action in the nonabelian case. In fact, it is not usually continuous in the g variable alone. Hence, if for $x \in X$ we construct a point $p \in S$ such that $h^{-1}phx = x$ for all $h \in G_0$ it need not follow that the equation holds for all $h \in G$.

Theorem 6.7. Let (S, G, S^*) be a classical Ellis semigroup with G a countable group and let $\Phi : S \times X \to X$ be a classical action with X metrizable. The sets $\{x \in X : Cent(\Phi, x)x = \mathcal{N}_{adj}\Phi(x)\}$ and $\{x \in X : (S^* \cap Cent(\Phi, x))x = \mathcal{N}^*_{adj}\Phi(x)\}$ are dense, G_δ subsets of X.

PROOF. A countable group G is discrete. S^* is a G_δ and so $k\mathcal{B}$ is a countably generated filter, as is $k\mathcal{P}_+ = \{G\}$. In fact, when $S^* = S \setminus G$, $k\mathcal{B}$ is the countable filter of cofinite sets. Otherwise, $k\mathcal{B} = k\mathcal{P}_+$.

Since X is metrizable it admits a countable opene base $\mathcal{T}$. Apply (6.23) with $\mathcal{F} = \mathcal{P}_+$ and $= \mathcal{B}$. Since the boundary of an open set is a closed, nowhere dense set, the result follows from the Baire Category Theorem. □

We would of course like to extend this result from a countable discrete group G to a general separable group G. The obvious attempt fails for the reason described in the Remark following Lemma 6.6. Thus, the results associated with the following theorem are only useful when G is countable. The definitions of adjoint recurrent point and adjoint transitive point are completely general but only in the countable case can we show that such points exist. See Theorem 6.8(d).

For $\Phi : S \times X \to X$ a classical Ellis action, $x \in X$ and $g \in G$, observe that

(6.25) $$gIso_x g^{-1} = Iso_{gx}.$$

We define the *adjoint isotropy set*
(6.26)
$$Cent(\Phi, x) \cap Iso_x = \bigcap_{g \in G} gIso_x g^{-1} = \bigcap_{g \in G} Iso_{gx} =_{def} Iso_x^{adj}.$$

As the intersection of closed co-ideals, this is a closed co-ideal which always contains the identity e. By definition $p \in Cent(\Phi, x) \cap Iso_x$ iff $pgx = gx$ for all $g \in G$. We will call x an *adjoint recurrent point* when $Cent(\Phi, x) \cap Iso_x \cap S^* \neq \emptyset$. We denote by $RECUR^{adj}$ the set of adjoint recurrent points. The Ellis-Namakura Lemma implies that if x is an adjoint recurrent point then $Iso_x^{adj} \cap S^*$ contains idempotents.

We call a point $x \in X$ an *adjoint transitive point* if $Cent(\Phi, x)x = X$, i.e. for every $y \in X$ there exists $p \in S$ such that $pgx = gy$ for all $g \in G$. If p can always be chosen in S^* then we call x an S^* *adjoint transitive point*. Thus, x is an S^* adjoint transitive point when $(S^* \cap Cent(\Phi, x))x = X$. We let $TRANS^{adj}$ and $TRANS_{S^*}^{adj}$ denote the set of adjoint transitive points and the set of S^* adjoint transitive points.

We call Φ *adjoint point transitive* when $TRANS^{adj} \neq \emptyset$ and S^* *adjoint point transitive* when $TRANS_{S^*}^{adj} \neq \emptyset$. Of course, if $S^* = S$ then all of the S^* concepts are the same as the originals.

Theorem 6.8. Let (S, G, S^*) be a classical Ellis semigroup with G a group and let $\Phi : S \times X \to X$ be a classical action.

(a) $RECUR^{adj}$ is a G invariant subset of X.

(b) If Φ is adjoint point transitive then $TRANS^{adj}$ is a dense, G invariant subset of X and Φ is adjoint transitive.

(c) A point is an S^* adjoint transitive point iff it is an adjoint recurrent, adjoint transitive point, i.e.

(6.27) $$TRANS_{S^*}^{adj} = TRANS^{adj} \cap RECUR^{adj}.$$

If Φ is S^* adjoint point transitive then $TRANS_{S^*}^{adj} = TRANS^{adj}$ and so is a dense, G invariant subset of X. Furthermore, Φ is then S^* adjoint transitive.

(d) Assume that G is a countable group and that X is metrizable. If Φ is adjoint transitive then $TRANS^{adj}$ is a dense, G_δ, G invariant subset of X and Φ is adjoint point transitive. If Φ is S^* adjoint transitive then $TRANS^{adj}_{S^*}$ is a dense, G_δ, G invariant subset of X and Φ is S^* adjoint point transitive.

PROOF. (a): The condition that $p \in S^*$ fixes gx for every $g \in G$ depends only on the G orbit of x.

(b),(c): Assume $x \in TRANS^{adj}$, $y \in X$ and $g \in G$. Because x is an adjoint transitive point, there exist $p, q \in S$ such that $phx = hy$ for all $h \in G$ and such that $qh_1 x = h_1 g^{-1} x$ for all $h_1 \in G$. Then with $h_1 = hg$ we see that for all $h \in G$

$$(6.28) \qquad (pq)h(gx) \;=\; phgg^{-1}x \;=\; phx \;=\; hy$$

That is, pq is the required element of S which maps each translate of gx to the translate of y. It follows that $TRANS^{adj}$ is G invariant. It is contained in $TRANS$ and so Gx is dense if $x \in TRANS^{adj}$. Thus, $TRANS^{adj}$ is dense if it is nonempty.

If $x \in TRANS^{adj}_{S^*}$ then there exists an idempotent $u \in S^*$ such that $ugx = gx$ for all $g \in G$ and so $x \in RECUR^{adj}$. Hence, $TRANS^{adj}_{S^*}$ is contained in $TRANS^{adj} \cap RECUR^{adj}$. Now we assume that there exists $x \in TRANS^{adj} \cap RECUR^{adj}$ and $z \in TRANS^{adj}$. We show that $z \in TRANS^{adj}_{S^*}$. This will show that equation (6.27) holds and that $TRANS^{adj} = TRANS^{adj}_{S^*}$ when the latter is nonempty. Given $y \in X$ there exist $p, q \in S$ and $r \in S^*$ such that for all $g \in G$: $pgx = gy$, $qgz = gx$ and $rgx = gx$. It follows that $(prq)gz = gy$ for all $g \in G$. Since $SS^*S \subset S^*$ it follows that $prq \in S^*$. If Φ is S^* adjoint point transitive then $TRANS^{adj} = TRANS^{adj}_{S^*}$ and so the latter is G invariant and dense. From (6.24) we have

$$(6.29) \qquad \begin{aligned} Cent(\Phi, x)x &\subset \mathcal{N}_{adj}\Phi(x) \\ (Cent(\Phi, x) \cap S^*)x &\subset \mathcal{N}^*_{adj}\Phi(x) \end{aligned}$$

So $\mathcal{N}_{adj}\Phi(x) = X$ if x is an adjoint transitive point. Since the set of adjoint transitive points is either empty or dense and $\mathcal{N}_{adj}\Phi$ is a closed relation it follows that adjoint point transitivity implies point transitivity. Similarly, for S^* adjoint transitivity.

(e): Apply Theorem 6.7. If Φ is adjoint transitive then $TRANS^{adj} = \{x \in X : Cent(\Phi, x)x = \mathcal{N}_{adj}\Phi(x)\}$. Similarly, when Φ is S^* adjoint transitive then $TRANS^{adj}_{S^*} = \{x \in X : (Cent(\Phi, x) \cap S^*)x = \mathcal{N}^*_{adj}\Phi(x)\}$. □

We now emerge from the straightjacket of countability.

It will be helpful to describe the different versions of transitivity by using family language.

For a system Φ we define the families on G:

(6.30)
$$\mathcal{J}^\Phi =_{def} [[\{N^\Phi(U,V) : \text{for opene} \quad U, V \subset G \}]].$$
$$\overline{\mathcal{J}}^\Phi =_{def} \mathcal{J}^\Phi \cdot k\mathcal{B}.$$
$$\mathcal{J}^\Phi_{adj} =_{def} [[\{N^\Phi_F(U,V) : \text{for opene } U, V \subset G \text{ and finite } F \subset G \}]].$$
$$\overline{\mathcal{J}}^\Phi_{adj} =_{def} \mathcal{J}^\Phi_{adj} \cdot k\mathcal{B}.$$

Recall that a family is proper when it does not contain the empty set. Clearly, Φ is transitive (or S^* transitive) iff $\mathcal{J}^\Phi$ is a proper family (resp. iff $\overline{\mathcal{J}}^\Phi$ is a proper family). Φ is adjoint transitive (or S^* adjoint transitive) iff $\mathcal{J}^\Phi_{adj}$ is a proper family (resp. iff $\overline{\mathcal{J}}^\Phi_{adj}$ is a proper family).

Recall that Φ is called *weak mixing* when the product action Φ^2 on $X \times X$ is transitive.

Lemma 6.9. Let (S, G, S^*) be a classical Ellis semigroup with G a group and let $\Phi : S \times X \to X$ be a classical action.

(a) If *either*: for all U, V opene subsets of X and $g \in G$,

(6.31)
$$N^\Phi(U,V) \cap N^\Phi(gU, gU) \neq \emptyset,$$

or: for all U, V opene subsets of X and $g \in G$,

(6.32)
$$N^\Phi(V,U) \cap N^\Phi(gU, gU) \neq \emptyset,$$

then Φ is weak mixing.

(b) If Φ is weak mixing then it is S^* transitive.

(c) Assume that Φ is adjoint transitive. If *either*: for all U, V opene subsets of X ,

(6.33)
$$N^\Phi(U,V) \cap N^\Phi(U,U) \neq \emptyset,$$

or: for all U, V opene subsets of X ,

(6.34)
$$N^\Phi(V,U) \cap N^\Phi(U,U) \neq \emptyset,$$

then Φ is weak mixing.

PROOF. Note first that (6.8) implies that $(N(U,V) \cap N(gU, gU))^{-1} = N(V,U) \cap N(gU, gU)$. So in the group case assumptions (6.31) and (6.32) are equivalent as are (6.33) and (6.34).

(a): By (6.31), $N(U, V)$ is nonempty for all opene U, V and so Φ is transitive.

Suppose we are given opene U_1, V_1, U_2, V_2 subsets of X. Clearly,

(6.35) $\quad N^{\Phi \times \Phi}(U_1 \times V_1, U_2 \times V_2) = N^\Phi(U_1, U_2) \cap N^\Phi(V_1, V_2).$

For weak mixing we must show that this set is nonempty.

There exist $g_1, g_2 \in G$ such that $U = U_1 \cap g_1^{-1} V_1 \cap g_2^{-1} U_2$ is opene because Φ is transitive. Let $V = g_1^{-1} g_2^{-1} V_2$ and let $g = g_1^{-1}$. From (6.8) we see that

(6.36)
$$\begin{aligned} N(gU, gU) \cap N(U,V) &\subset N(g_1^{-1} U_1, g_1^{-1} g_2^{-1} U_2) \cap N(g_1^{-1} V_1, g_1^{-1} g_2^{-1} V_2) \\ &= g_1^{-1} g_2^{-1} (N(U_1, U_2) \cap N(V_1, V_2)) g_1. \end{aligned}$$

From assumption (6.31) it follows that $N^\Phi(U_1, U_2) \cap N^\Phi(V_1, V_2)$ is nonempty, as required.

(b): We apply Lemma 5.29. Observe first that a classical Ellis semigroup with G a group admits no bizarre actions. By definition a bizarre action on X admits a point $x \in X$ such that $Gx = X$. Since G is a group, X must be minimal. However, a bizarre action on X has a proper closed invariant subset X^* and so X cannot be minimal. If Φ is not S^* transitive then it falls either into Case (0) or Case (3) of Theorem 6.3 and so it is not S^* point transitive either. Lemma 5.29 then implies that Φ is not weak mixing. Our result is the contrapositive.

(c): We use adjoint transitivity to get (6.31) from (6.33). Given opene U, V and $g \in G$, we can use (6.33) to choose $h \in G$ such that $U \cap h^{-1} U$ and $U \cap h^{-1} V$ are opene. Now let $F = \{g, h\}$. Observe that

(6.37) $\qquad N_F(U \cap h^{-1} U, U \cap h^{-1} V) \;\subset\; N(gU, gU) \cap N(U,V).$

These are nonempty by the assumption of adjoint transitivity. Applying part (a), we obtain weak mixing. $\qquad\square$

Using adjoint transitivity we obtain the nonabelian *Furstenberg Intersection Lemma*:

Theorem 6.10. Let (S, G, S^*) be a classical Ellis semigroup with G a group. Let $\Phi : S \times X \to X$ be an adjoint transitive classical action which is weak mixing. For all U_1, V_1, U_2, V_2 opene subsets of X and F_1, F_2 finite subsets of G, there exist U_3, V_3 opene subsets of X and F_3 a finite subset of G such that

(6.38) $\qquad N^\Phi_{F_3}(U_3, V_3) \;\subset\; N^\Phi_{F_1}(U_1, V_1) \cap N^\Phi_{F_2}(U_2, V_2).$

PROOF. From the assumption of weak mixing we can choose $h \in N^\Phi(U_1, U_2) \cap N^\Phi(V_1, V_2)$. Let $U_3 = U_1 \cap h^{-1} U_2$ and $V_3 = V_1 \cap h^{-1} V_2$, opene subsets by choice of h. Define $F_3 = F_1 \cup F_2 h$. We have

(6.39)
$$\begin{aligned} N_{F_3}(U_3, V_3) &\subset N_{F_1}(U_1, V_1) \cap N_{F_2 h}(h^{-1} U_2, h^{-1} V_2) \\ &= N_{F_1}(U_1, V_1) \cap N_{F_2}(U_2, V_2). \end{aligned}$$

$\qquad\square$

We will call the classical action $\Phi : S \times X \to X$ *adjoint weak mixing* when the product action $\Phi^2 : S \times X \times X \to X \times X$ is adjoint transitive.

Theorem 6.11. Let (S, G, S^*) be a classical Ellis semigroup with G a group and let $\Phi : S \times X \to X$ be a classical action.
   (a) The following conditions on Φ are equivalent.
       (1) Φ is adjoint weak mixing.
       (2) Φ is adjoint transitive and weak mixing.
       (3) For every positive integer n the product action $\Phi^n : S \times X^n \to X^n$ is S^* adjoint transitive.
       (4) For every positive integer n the product action $\Phi^n : S \times X^n \to X^n$ is transitive, i.e. Φ is weak mixing of all orders.
       (5) The family $\mathcal{T}^\Phi_{adj}$ is a filter, i.e. it is closed under intersection.
       (6) The family $\overline{\mathcal{T}}^\Phi_{adj}$ is a filter.
       (7) There exists a filter $\mathcal{F}$ of subsets of G such that $\mathcal{T}^\Phi \subset \mathcal{F}$.
   (b) Adjoint weak mixing is a residual property stronger than S^* adjoint transitivity.
   (c) If Φ is adjoint weak mixing then it is scattering, and for every index set I the product system $\Phi^I : S \times X^I \to X^I$ is adjoint weak mixing and so is S^* adjoint transitive.
   (d) If Φ is mixing, i.e. $k\mathcal{B}$ transitive, then it is adjoint weak mixing.

PROOF. (a), (1) $\Rightarrow$ (2): Obvious.
(2) $\Rightarrow$ (5): This follows from the Furstenberg Intersection Lemma, Theorem 6.10.
(5) $\Rightarrow$ (7): Obvious.
(7) $\Rightarrow$ (4): This follows from the obvious extension of (6.35):

$$(6.40) \qquad N^{\Phi^n}(U_1 \times ... \times U_n, V_1 \times ... \times V_n) = \bigcap_{i=1}^{n} N^\Phi(U_i, V_i).$$

(4) $\Rightarrow$ (3): Since each Φ^{2n} is transitive, it follows from Lemma 6.9(b) that each Φ^n is S^* transitive. Each $N_F(U,V)$ is just a finite intersection of $N(U_i, V_i)$'s. It follows that each Φ^n is S^* adjoint transitive.
(3) $\Rightarrow$ (1): Obvious. This shows that all but (6) are equivalent to one another.
(6) $\Rightarrow$ (7): Obvious.
(3) $\Rightarrow$ (6): Condition (3) implies that $\mathcal{T}^\Phi_{adj} \subset \mathcal{B}$ and so $\mathcal{T}^\Phi_{adj} \cdot k\mathcal{B}$ is a proper family. Since (3) implies (5), $\mathcal{T}^\Phi_{adj}$ is a filter. Since $k\mathcal{B}$ is a filter as well, it follows that $\overline{\mathcal{T}}^\Phi_{adj} = \mathcal{T}^\Phi_{adj} \cdot k\mathcal{B}$ is closed under intersection and is a proper family, i.e. it is also a filter.
(b): Since adjoint transitivity is residual, this follows from Proposition 5.24(b). Note that (1) implies (3) in (a) shows that adjoint weak mixing is stronger than S^* adjoint transitivity.

(c): If $\mathfrak{T}^\Phi \subset \mathfrak{F}$ it is easy to check that $\mathfrak{T}^{\Phi^I} \subset \mathfrak{F}$ for any index set I. Hence, Φ^I is adjoint weak mixing when Φ is.

Now assume that $\Psi : S \times Y \to Y$ is a minimal system with W_1, W_2 opene subsets of Y. Let U_1, U_2 be opene subsets of X. Because Y is minimal, there exists a finite $F \subset G$ such that

(6.41)
$$\bigcup_{g \in F} gW_2 = Y$$

and so
$$\bigcup_{g \in F} N(W_1, gW_2) = G.$$

Since the intersection of the $N^\Phi(U_1, gU_2)$'s is nonempty, it follows that for some $g \in F$ there exists $k \in N^\Phi(U_1, gU_2) \cap N(W_1, gW_2)$ and so $g^{-1}k \in N^\Phi(U_1, U_2) \cap N(W_1, W_2)$.

Hence, $\Phi \times \Psi$ is transitive.

(d): This follows from condition (7) of part (a), because $k\mathcal{B}$ is a filter. □

REMARK 6.5. Because Φ is S^* weak mixing of all orders, there exists $A \in \mathcal{B}$ which is contained in $N^\Phi(U_1, gU_2)$ for all $g \in F$. Then (6.41) implies that

(6.42)
$$\bigcup_{g \in F} N^\Phi(U_1, gU_2) \cap N(W_1, gW_2) \supset \bigcup_{g \in F} A \cap N^\Phi(U_1, gU_2) = A \in \mathcal{B}.$$

which implies that $N^\Phi(U_1, gU_2) \cap N(W_1, gW_2) \in \mathcal{B}$ for some $g \in F$ because $\mathcal{B}$ is a filterdual. This implies that the product action $\Phi \times \Psi$ is S^* transitive.

Example (6) The "$2 \Rightarrow n$" result which comes from the Furstenberg Intersection Lemma does not hold for ordinary weak mixing in the nonabelian case. For example, let X be the circle and G be the entire group of homeomorphisms on X which acts by evaluation on X. Any triple of distinct points of X can be mapped to any other by a homeomorphism. Hence, the action of G on X^3 is topologically transitive and so the action on X is weak mixing. On the other hand, define U_1 (and U_2) to be those quadruples (x_1, x_2, x_3, x_4) where x_3 and x_4 lie in the same component of $X \setminus \{x_1, x_2\}$ (respectively, in different components of $X \setminus \{x_1, x_2\}$). Then U_1 and U_2 are disjoint G invariant opene subsets of X^4. Hence, the action on X^4 is not topologically transitive. Notice also that the action of G on X is proximal, i.e. any pair of points in X is a proximal pair. Hence, $X \times X = PROX = LPROX$.

There is an example, given in Peleg (1972) of a countable subgroup of G such that the restricted action is still minimal on X and transitive on X^3. See also Weiss (2000) and Glasner (2005). The latter contains a detailed discussion of the various notions of weak mixing in the nonabelian case.

On the other hand, the analogous result that weak mixing implies weak mixing of all orders is true in the measure theory category. We obtain the following theorem whose proof we merely sketch.

Theorem 6.12. Let (S, G, S^*) be a classical Ellis semigroup with G a group and let $\Phi : S \times X \to X$ be a classical action. Assume there exists a Borel probability measure μ on X, invariant with respect to the G action, and such that $\mu(U) > 0$ for all opene $U \subset X$, i.e. μ has *full support*. If the product measure $\mu \times \mu$ on $X \times X$ is ergodic with respect to the G action on $X \times X$, then Φ is adjoint weak mixing.

PROOF. Ergodicity of an invariant measure of full support implies transitivity of the G action, i.e. $N(U, V) \neq \emptyset$ for all opene $U, V \subset X$. On the other hand, if $\mu \times \mu$ is ergodic on $X \times X$ then μ^n is ergodic on X^n, i.e. the "2 $\Rightarrow$ n" property holds. Since Φ^n is then transitive for all positive integers n, it follows from Theorem 6.11(a) that Φ is adjoint weak mixing. See Glasner (2003) Chapter 3. □

In the minimal case, we can relate proximality and weak mixing.

Theorem 6.13. Let (S, G, S^*) be a classical Ellis semigroup with G a separable group and let $\Phi : S \times X \to X$ be an adjoint transitive classical action with X minimal. The following conditions are equivalent

(1) Φ is adjoint weak mixing.
(2) Φ is weak mixing.
(3) $N^\Phi(U_1, V) \cap ... \cap N^\Phi(U_n, V) \neq \emptyset$ for all opene $U_1, ..., U_n, V \subset X$.
(4) The regional proximality relation is the entire space, i.e. $QPROX(\Phi) = X \times X$.
(5) The proximality relation $PROX(\Phi)$ is dense in $X \times X$.

PROOF. (1) ⇔ (2): Apply Theorem 6.11(a).
(1) ⇒ (3): By Theorem 6.11(a), Φ^n is transitive.
(3) ⇒ (4): Obvious.
(4) ⇒ (5): Apply Proposition 5.16(d).
(5) ⇒ (2): Given opene $U, V \subset X$, there exists $(x, y) \in (V \times U) \cap PROX$. So there exists $z \in X$ and $p \in S$ such that $px = z = py$. Because X is minimal, there exists $g \in G$ such that $gz \in U$. Since $gpx = gz = gpy \in U$ there exists $g_1 \in G$ close enough to gp that $g_1 x, g_1 y \in U$. Thus, Φ satisfies condition (6.34). By Lemma 6.9(c), Φ is weak mixing. □

There is a useful map result related to Theorem 6.11. Suppose we are given classical actions $\Phi : S \times X \to X, \Psi : S \times Y \to Y, \Theta : S \times Z \to Z$ and surjective, continuous action maps $\pi : X \to Z, \rho : Y \to Z$. The *pullback* $R_{\pi,\rho}$ and the map $\lambda_{\pi,\rho} : R_{\pi,\rho} \to Z$ are given by

(6.43)
$$R_{\pi,\rho} =_{def} \{(x,y) \in X \times Y : \pi(x) = \rho(y)\} = (\pi \times \rho)^{-1}(1_Z).$$
$$\lambda_{\pi,\rho}(x,y) =_{def} \pi(x) = \rho(y).$$

$R_{\pi,\rho}$ is a closed, invariant subset of $X \times Y$ and so the product action $\Phi \times \Psi$ restricts to a classical Ellis action on the pullback and $\lambda_{\pi,\rho}$ is a continuous action map. Notice that if $\pi = \rho$ then the pullback is just R_π.

We need a technical result first.

Lemma 6.14. Let (S, G, S^*) be a classical Ellis semigroup with G a group. Assume that we are given classical actions $\Phi : S \times X \to X, \Psi : S \times Y \to Y, \Theta : S \times Z \to Z$ and surjective, continuous action maps $\pi : X \to Z, \rho : Y \to Z$. Assume that Y and hence Z are minimal and that π is an open map.

(a) If A is an opene subset of Y then $\rho(A)$ has a nonempty interior in Z. Furthermore, the interior of $\rho(A)$ is dense in $\rho(A)$.

(b) If D is an open subset of $X \times Y$ which meets $R_{\pi,\rho}$ then there exist opene subsets $U \subset X$ and $A \subset Y$ such that $U \times A \subset D$ and $\pi(U) = \rho(A)$.

(c) If ρ, as well as π, is an open map then $\lambda_{\pi,\rho}$ is an open map.

PROOF. (a): Let K be a closed subset of A with nonempty interior. Because Y is minimal there is a finite subset $F \subset G$ such that $\{gK : g \in F\}$ covers Y and hence $\{\rho(gK) = g\rho(K) : g \in F\}$ is a finite cover of Z. Since the sets $\rho(gK)$ are closed, at least one of them has nonempty interior. Since G acts by homeomorphisms they all do. Now if $y \in A$ and B is an open subset of Z containing $\rho(y)$ then $A \cap (\rho)^{-1}(B)$ is an opene subset of Y and so its image has nonempty interior. Thus, the interior of $\rho(A)$ meets any open set B which meets $\rho(A)$.

(b): Notice that $U \times A \subset X \times Y$ meets $R_{\pi,\rho}$ iff $\pi(U)$ meets $\rho(A)$. First, choose opene sets $U_0 \subset X$ and $A_0 \subset Y$ such that $U_0 \times A_0 \subset D$ and $U_0 \times A_0$ meets $R_{\pi,\rho}$. By assumption $\pi(U_0)$ is open and $\rho(A_0)$ meets $\pi(U_0)$. By part (a), the interior, $Int\rho(A_0)$ is dense in $\rho(A_0)$ and so meets $\pi(U_0)$ as well. Let $W = \pi(U_0) \cap Int\rho(A_0), U = U_0 \cap \pi^{-1}(W)$ and $A = A_0 \cap \rho^{-1}W$.

(c): Proceed as in (b), but with an easier argument since ρ is open. If D is open with $(x, y) \in D \cap R_{\pi,\rho}$ then we can choose open subsets $U_0 \subset X$ and $A_0 \subset Y$ such that $U_0 \times A_0 \subset D$ and $x \in U_0, y \in A_0$. Let V be the open set $\pi(U_0) \cap \rho(A_0)$. For any $z \in V$ there exist $x_1 \in U_0, y_1 \in A_0$ such that $\pi(x_1) = z\rho(y_1)$. Since $(x_1, y_1) \in D \cap R_{\pi,\rho}$ it follows that $\lambda_{\pi,\rho}(D)$ contains V. Thus, $\lambda_{\pi,\rho}$ is an open map. $\square$

The following Theorem as well as Theorems 6.24, 6.25 and the ensuing Corollaries 6.26 and 6.27 extend results of Glasner (1967) (Theorem II.2.1) and van der Woude (1986) (see also Glasner (2005)).

Theorem 6.15. Let (S, G, S^*) be a classical Ellis semigroup with G a group. Assume that we are given classical actions $\Phi : S \times X \to X, \Psi : S \times Y \to Y, \Theta : S \times Z \to Z$ and surjective, continuous action maps $\pi : X \to Z, \rho : Y \to Z$ with Y and hence Z minimal and with π an open map. Suppose there exists $z \in Z$ such that the fiber $X_z = \pi^{-1}\{z\}$ satisfies the following condition: Whenever $U_1, ..., U_n$ are open subsets of X which meet X_z then for all opene $V \subset X$

(6.44) $\qquad N^\Phi(U_1 \cap X_z, V) \cap .... \cap N^\Phi(U_n \cap X_z, V) \neq \emptyset.$

Then the restriction $\Phi \times \Psi | R_{\pi,\rho}$ is transitive. If each intersection in (6.44) is a member of $\mathcal{B}$ (see (5.28)), then the restriction $\Phi \times \Psi | R_{\pi,\rho}$ is S^* transitive.

PROOF. We must show that given relatively open, nonempty subsets $D, E \subset R_{\pi,\rho}$, the hitting time set $N(D, E)$ is nonempty for the product action. By Lemma 6.14 (b), we can assume $D = (U \times A) \cap R_{\pi,\rho}$ and $E = (V \times B) \cap R_{\pi,\rho}$ where $U, V \subset X$, $A, B \subset Y$ are opene with $\pi(U) = \rho(A)$ and $\pi(V) = \rho(B)$. Because Y is minimal there is a finite subset $F \subset G$ such that $\{gA : g \in F\}$ covers Y. Discard those members of F such that gA is disjoint from $Y_z = \rho^{-1}\{z\}$. We are left with a finite subset F such that $\{gA : g \in F\}$ covers Y_z and each member meets Y_z. Hence, we have:

(6.45) $\qquad z \;\in\; \rho(gA) \;=\; g\rho(A) \;=\; g\pi(U) \;=\; \pi(gU),$

for all $g \in F$.

By assumption (6.44) there exists $h \in \bigcap_{g \in F} N(X_z \cap gU, V)$. Hence, we have

(6.46) $\qquad hz \;\in\; \pi(V) \;=\; \rho(B).$

Since $h^{-1}B$ meets Y_z it meets $g_0 A$ for some $g_0 \in F$ and so $h \in N(Y_z \cap g_0 A, B)$. Let $w = g_0^{-1} z$. We have $hg_0 \in N((U \times A) \cap (X_w \times Y_w), V \times B)$ as required.

The proof actually shows that
(6.47)
$$\bigcap_{g \in F} N(X_z \cap gU, V) \;\subset\; \bigcup_{g \in F} N((U \times A) \cap (X_{g^{-1}z} \times Y_{g^{-1}z}), V \times B)g^{-1}.$$

It follows that if each intersection in (6.44) is a member of $\mathcal{B}$, then one of the sets in the above union lies in the filterdual $\mathcal{B}$ by the Ramsey Property. Since S^* is a two-sided ideal by (5.3), the family $\mathcal{B}$ is invariant by left and right translation of elements of G. It follows that the restriction $\Phi \times \Psi | R_{\pi,\rho}$ is S^* transitive. $\qquad\square$

From this we obtain a sharpening of Theorem 6.11(c). See Theorem 6.13, as well.

Corollary 6.16. Let (S, G, S^*) be a classical Ellis semigroup with G a group. Assume a classical action $\Phi : S \times X \to X$ satisfies the condition:

(6.48) $$N^\Phi(U_1, V) \cap ... \cap N^\Phi(U_n, V) \neq \emptyset$$

for all opene $U_1, ..., U_n, V \subset X$, then Φ is scattering.

PROOF. Apply Theorem 6.15 to the special case when Θ is a trivial action. Then π is clearly open and the pullback is the product action on $X \times Y$. □

The conditions given in (6.44) and (6.48) are closely related to the n-fold version of proximality. This is an opportune moment to introduce this extension of the proximality concept.

If S is an Ellis semigroup, $\Phi : S \times X \to X, \Psi : S \times Y \to Y$ are Ellis actions and $\pi : X \to Y$ a continuous action map then for a positive integer n we denote by $\Phi^n : S \times X^n \to X^n$ and $\Psi^n : S \times Y^n \to Y^n$ the product actions and by $\pi^n : X^n \to Y^n$ the product action map. Define:

(6.49)
$$\Delta_X^n =_{def} \{(x_1, ..., x_n) \in X^n : x_1 = \cdots = x_n\}.$$
$$PROX^n(\Phi) =_{def} \{(x_1, ..., x_n) \in X^n : px_1 = \cdots = px_n \text{ for some } p \in S\}.$$
$$R_\pi^n =_{def} \{(x_1, ..., x_n) \in X^n : \pi(x_1) = \cdots = \pi(x_n)\} = (\pi^n)^{-1}(\Delta_Y^n).$$
$$\lambda_\pi^n : R_\pi^n \to Y \text{ by } \lambda_\pi^n(x_1, ..., x_n) =_{def} \pi(x_1) = \cdots = \pi(x_n).$$

Lemma 6.17. Let $\Phi : S \times X \to X$ be an Ellis action. Let $(x_1, ..., x_n) \in X^n$. If for $i = 1, ..., n-1$, $(x_i, x_{i+1}) \in LPROX(\Phi)$, then $(x_1, ..., x_n) \in PROX^n(\Phi)$.

PROOF. By (4.10) $LPROX = ASYMP_{[Min(S)]}$ and so $LPROX$ is an equivalence relation and on each equivalence class Φ^p is a constant map for every minimal p. □

Proposition 6.18. Let $\Phi : S \times X \to X$ and $\Psi : S \times Y \to Y$ be Ellis actions. If $\pi : X \to Y$ is a proximal, continuous action map then $R_\pi^n \subset PROX^n(\Phi)$.

PROOF. By Theorem 4.5 (c), $R_\pi \subset LPROX$ and so Φ^p is constant on every fiber of π for every minimal p. □

Now let (S, G, S^*) be a classical Ellis semigroup. Assume that $\Phi : S \times X \to X$ and $\Psi : S \times Y \to Y$ are classical actions with $\pi : X \to Y$ a continuous action map. Define

(6.50)
$$QPROX^n(\Phi) =_{def} (\mathcal{N}\Phi^n)^{-1}(\Delta_X^n) \subset X^n.$$
$$QPROX^n(\pi) =_{def} (\mathcal{N}(\Phi^n|R_\pi^n))^{-1}(\Delta_X^n) \subset R_\pi^n.$$

In particular, $PROX(\Phi)$ and $QPROX(\Phi)$ defined earlier are the special cases with $n = 2$.

For a classical action $\Phi : S \times X \to X$ of a classical Ellis semigroup (S, G, S^*) and $A \subset X$ we will use the notation.

(6.51) $\quad G^{-1}A =_{def} \bigcup\{(\Phi^g)^{-1}(A) : g \in G\}.$

Proposition 6.19. Let (S, G, S^*) be a classical Ellis semigroup. Assume that we are given classical actions $\Phi : S \times X \to X, \Psi : S \times Y \to Y$ and $\pi : X \to Y$ a continuous action map. Let n be a positive integer and let $\mathcal{N}^n$ denote the filter of neighborhoods of Δ^n in X^n.

(6.52) $\quad PROX^n(\Phi) = \bigcap\{G^{-1}V : V \in \mathcal{N}^n\},$

(6.53) $\quad QPROX^n(\Phi) = \bigcap\{\overline{G^{-1}V} : V \in \mathcal{N}^n\},$

(6.54) $\quad QPROX^n(\pi) = \bigcap\{\overline{R_\pi^n \cap G^{-1}V} : V \in \mathcal{N}^n\},$

PROOF. A point $z = (x_1, ..., x_n) \in X^n$ is in $PROX^n$ iff $pz \in \Delta^n$ for some $p \in S$. Since G is dense in S and Φ_z^n is continuous this occurs iff for every neighborhood V of Δ^n, $gz \in V$ for some $g \in G$. On the other hand, $z \in QPROX$ iff for every such V and every neighborhood U of z there exists $\tilde{z} \in U$ and $g \in G$ such that $g\tilde{z} \in V$. Finally, if $z \in R_\pi^n$ then $z \in QPROX^n(\pi)$ iff, in addition, the perturbed point $\tilde{z}$ can always be chosen to lie in $U \cap R_\pi^n$. □

From this Proposition we clearly get:

(6.55)
$$PROX^n(\Phi) \subset QPROX^n(\Phi) \quad \text{and}$$
$$R_\pi^n \cap PROX^n(\Phi) \subset QPROX^n(\pi) \subset R_\pi^n \cap QPROX^n(\Phi)$$

Corollary 6.20. Let (S, G, S^*) be a classical Ellis semigroup. Assume that $\Phi : S \times X \to X$ is a classical action and that n is a positive integer. If X is metrizable then $PROX^n(\Phi)$ is a G_δ subset of X^n and it is dense in X^n iff $QPROX^n(\Phi) = X^n$.

If, in addition, $\Psi : S \times Y \to Y$ is a classical action and $\pi : X \to Y$ is a continuous action map, then the G_δ set $R_\pi^n \cap PROX^n(\Phi)$ is dense in R_π^n iff $QPROX^n(\pi) = R_\pi^n$.

PROOF. Since X is a compact metric space, the filter $\mathcal{N}^n$ of neighborhoods of Δ^n has a countable base consisting of open sets. Intersecting over these we see from (6.52) that $PROX^n$ is a G_δ.

By (6.54) $QPROX^n(\pi) = R_\pi^n$ says exactly that for every $V \in \mathcal{N}^n$, $R_\pi^n \cap G^{-1}V$ is a dense open subset of R_π^n. So the Baire Category Theorem implies that $R_\pi^n \cap PROX^n$ is dense in R_π^n. The converse is obvious because $QPROX^n(\pi)$ is closed and contains $R_\pi^n \cap PROX^n$.

The original space result is the special case which uses the action map of X to the trivial action.

$\square$

When the action map π is open we can relate the conditions on $QPROX$ to those of (6.44) and (6.48). The proof will require a topological version of Fubini's Theorem. For its proof see Veech(1970) Proposition 3.1, Glasner (1990) Lemma 5.2 and the Remark thereafter, or Akin (2004) Theorem 1.2 (see also the Kuratowski-Ulam Theorem in Oxtoby (1980) Section 15).

Theorem 6.21. Let $\pi : X \to Y$ be a continuous open surjection with X metrizable. If D is a dense G_δ subset of X then $\{y \in Y : $ the G_δ set $D \cap \pi^{-1}(y)$ is dense in $\pi^{-1}(y)\}$ is a dense G_δ subset of Y.

REMARK 6.6. Notice that if $\pi : X \to Y$ is a surjection of compact spaces and X is metrizable, then Y is metrizable as well, as it is easy to see that second countability is preserved.

Proposition 6.22. Let (S, G, S^*) be a classical Ellis semigroup. Assume that $\Phi : S \times X \to X$ and $\Psi : S \times Y \to Y$ are classical actions and that $\pi : X \to Y$ is a continuous action map. For $y \in Y$ let X_y denote the fiber $\pi^{-1}(y)$. If X is minimal and π is an open map then the following conditions are equivalent.

(1) There exists $y \in Y$ such that whenever $U_1, ..., U_n, V$ are open subsets of X which meet X_y

(6.56) $\quad N^\Phi(U_1 \cap X_y, V) \cap .... \cap N^\Phi(U_n \cap X_y, V) \neq \emptyset.$

(2) For every $y \in Y$ whenever $U_1, ..., U_n$ are open subsets of X which meet X_y then for all opene $V \subset X$ (6.56) holds.

(3) $QPROX^n(\Phi) = R_\pi^n$.

If, in addition, X is metrizable then these conditions imply that there exists a dense G_δ subset D of Y such that $PROX^n(\Phi) \cap X_y^n$ is dense in X_y^n for all $y \in D$.

PROOF. Notice first that the map λ_π^n is open when π is open. The argument in Lemma 6.14(c) for the $n = 2$ case easily generalizes.

(3) $\Leftrightarrow$ (2): Since Y is minimal, $Q = G^{-1}(V^n)$ is a neighborhood of Δ^n. Condition (3) implies there exists $z = (x_1, ..., x_n) \in (U_1 \times ... \times U_n) \cap R_\pi^n$ and a $g \in G$ such that $gz \in Q$ and so $g_1 g z \in V^n$ for some $g_1 \in G$. Hence, $g_1 g$ is a point in the intersection described in (6.56).

(2) $\Rightarrow$ (1): Obvious.

(1) $\Rightarrow$ (3): Let $Z = \{y \in Y : X_y^n \subset QPROX^n(\pi)\}$. It is easy to see that this is a G invariant subset of Y. Condition (1) implies that Z is nonempty. Since Y is minimal it follows that Z is dense. Let $V \in \mathcal{N}^n$ and let O be an opene subset of R_π^n. Since $\lambda_\pi^n(O)$ is opene it meets Z. Hence, $N(O, V)$ is nonempty. Since this is true for all opene O and all $V \in \mathcal{N}^n$, $QPROX(\pi) = R_\pi^n$.

When X is metrizable, the conditions imply that $R_\pi^n \cap PROX^n(\Phi)$ is dense in R_π^n by Corollary 6.20. Apply the Fubini Theorem 6.21 to this set and the opene map λ_π^n to obtain the set D. □

Some of our arguments will use the space 2^X, the set of closed subsets of X, equipped with the exponential topology, see Kuratowski (1966) Sections 17 and 42, see also Akin (2004) Section 4. It is a compact space, metrizable via the Hausdorff metric when X is a metric space. We will need the following properties

- The map $i_{X^n} : X^n \to 2^X$ by $(x_1, ..., x_n) \mapsto \{x_1, ..., x_n\}$ is continuous and is an embedding when n + 1. It follows that the sets with cardinality at most n form a closed subset of 2^X.
- If $A \subset X$ is closed (or if A is open) then $\{K \in 2^X : K \subset A\}$ and $\{K \in 2^X : K \cap A \neq \emptyset\}$ are closed subsets (resp. open subsets) of 2^X.
- The map $\vee : 2^X \times 2^X \to 2^X$ defined by $(A, B) \mapsto A \cup B$ is continuous.
- For a continuous map $\pi : X \to Y$ the image map $A \mapsto \pi(A)$ defines a continuous map $\pi_* : 2^X \to 2^Y$ and this, in turn, defines a continuous map $* : \mathcal{C}(X, Y) \to \mathcal{C}(2^X, 2^Y)$ which is a homomorphism of topological semigroups when $X = Y$.

- For a continuous map $\pi : X \to Y$ the closed inverse relation π^{-1} defines the preimage map given by $y \mapsto \pi^{-1}(y)$. This map $\pi^{-1} : Y \to 2^X$ is continuous iff π is an open map.
- The element relation $\{(x, K) : x \in K\}$ is a closed subset of $X \times 2^X$.

Proposition 6.23. Let (S, G, S^*) be a classical Ellis semigroup with G a group. Assume that $\Phi : S \times X \to X$ and $\Psi : S \times Y \to Y$ are classical actions and that $\pi : X \to Y$ is a continuous action map. Assume that X is minimal and that π is an open map. If some fiber $\pi^{-1}(y)$ is a finite set then every fiber $\pi^{-1}(y)$ is finite and all of the fibers have the same number of elements. For some positive integer n, the map π is an n-fold covering space map. Finally, π is a distal action map.

PROOF. : Because π is open, the fiber map $\pi^{-1} : Y \to 2^X$ is continuous. Also, for any $(g, y) \in G \times Y$, ϕ^g restricts to a homeomorphism of $\pi^{-1}(y)$ to $\pi^{-1}(gy)$.

Assume that some fiber is finite. For any positive integer n the set $F_n =_{def} \{y : \text{the cardinality of } \pi^{-1}(y) \text{ is at most } n\}$ is closed by continuity of π^{-1} and it is invariant under the action of G. Let n be the smallest cardinality of any fiber. Since Y is minimal F_n is all of Y. By choice of n every fiber contains exactly n elements. For a fixed y choose pairwise disjoint open sets $U_1, ..., U_n$ to cover $\pi^{-1}(y)$. Since π is open we can choose an open neighborhood V of y whose closure is contained in $\bigcup\{\pi(U_i) : i = 1, ..., n\}$. It is easy to see that $\pi^{-1}(V)$ decomposes into n pairwise disjoint open sets each mapping homeomorphically onto V, namely $V_i = U_i \cap \pi^{-1}(V), i = 1, ..., n$. That is, π is an n-fold covering space map.

If (x_1, x_2) is a proximal pair in $\pi^{-1}(z)$ then there exists a sequence $\{g_j\}$ in G such that $\{g_j z\}$ converges to y and $\{g_j x_1\}$ and $\{g_j x_2\}$ both converge to the same point of the fiber $\pi^{-1}(y)$. So eventually both sequences lie in the same set V_i on which π is injective. Hence, eventually $g_j x_1 = g_j x_2$. Since the g_j's are homeomorphisms, $x_1 = x_2$. That is, π is a distal map.
$\square$

We use all this to obtain some corollaries of Theorem 6.15.

We begin with a version of Corollary 6.16. It is essentially Theorem II.2.1 of Glasner (1976)

Theorem 6.24. Let (S, G, S^*) be a classical Ellis semigroup with G a group. Assume that $\Phi : S \times X \to X$ is a classical action with X minimal. If the action satisfies the condition that

(6.57) $\qquad N^{\Phi}(U_1, V) \cap ... \cap N^{\Phi}(U_n, V) \neq \emptyset$

for all opene $U_1, ..., U_n, V \subset X$ or, equivalently, that for every positive integer n $QPROX^n(\Phi) = X^n$, then Φ is scattering and weak mixing.

PROOF. The equivalence between the two conditions follows from Proposition 6.22. That the action is scattering just follows from Corollary 6.16. A scattering action Φ with X minimal is weak mixing because we can let $Y = X$ in that case.

$\square$

REMARK 6.7. Notice that the condition (6.57) implies transitivity and is implied by adjoint weak mixing. In the abelian case it is exactly adjoint weak mixing by the Furstenberg Intersection Lemma. However, in the non-abelian case these conditions are strictly weaker than adjoint weak mixing, or, equivalently, weak mixing of all orders. Recall Example (6) with G the homeomorphism group of the circle X. We can give G the discrete topology and extend to the Stone-Čech compactification to obtain a classical action on X. The action is minimal, weak mixing and proximal but is not adjoint weak mixing. Since $X \times X = PROX = LPROX$ it follows from Lemma 6.17 that $PROX^n = X^n$ for every positive integer n.

We now give a version of Theorem 6.15 with $\pi = \rho$. However, we replace condition (6.44) with one which is easier to use.

Theorem 6.25. Let (S, G, S^*) be a classical Ellis semigroup with G a group. Assume that $\Phi : S \times X \to X$, and $\Psi : S \times Y \to Y$ are classical actions and that $\pi : X \to Y$ a continuous action map. Assume that X is minimal and that π is an open map. Suppose there exists a point $y \in Y$ such that with $X_y = \pi^{-1}(y)$ we have that for every positive integer n the intersection $X_y^n \cap PROX^n(\Phi)$ is dense in X_y^n. Then the restriction of Φ^2 to R_π is S^* transitive. If, in addition, G is separable then the recurrent points are dense in R_π.

PROOF. We apply Theorem 6.15 with $\pi = \rho$.

If X_y is a single point then by Proposition 6.23 every fiber is a singleton and so $R_\pi = 1_X$ and $\Phi^2|R_\pi$ is isomorphic to the minimal action Φ which is transitive by Proposition 5.18(d). Furthermore, every point is minimal.

Suppose now that X_y contains more than one point. We verify the condition (6.44) for X_y. We are given Given open sets $U_1, ..., U_n$ which meet X_y, and an opene $V \subset X$.

Choose $x_i \in U_i \cap X_y$ for $i = 1, ..., n$ and choose $x_{n+1} \in X_y$ distinct from at least one of these so that $(x_1, ..., x_{n+1}) \notin \Delta^{n+1}$. Since $PROX^{n+1} \cap X_y^{n+1}$ is dense in X_y^{n+1} we can perturb the point so that we have, in addition, $(x_1, ..., x_{n+1}) \in PROX^{n+1}$. Since X is minimal there exists $p \in S$ such that $px_1 = \cdots = px_n = px_{n+1}$ and this common point lies in V. Since each element of the group G acts on X as a homeomorphism, while Φ^p is not injective, we have $p \in S \setminus G$. Hence, $Q = \{q \in S : qx_1, ..., qx_n \in V\}$ is an open subset of S which contains p and so meets $S \setminus G \supset S^*$. By Lemma 5.1(a) $N(x_1, V) \cap ... \cap N(x_{n+1}, V) = G \cap Q$ is a subset of G which is dense in Q. So it is an unbounded set in G and, so $N(x_1, V) \cap ... \cap N(x_{n+1}, V) \in \mathcal{B}$. By Theorem 6.15 the action on R_π is S^* transitive.

Finally, if G is separable, then Theorem 5.18(h) implies that the recurrent points are dense. Notice that G separable implies S^* is a G_δ by Proposition 6.1(d). □

Of course, if the map is proximal then it certainly satisfies the above condition that the proximal pairs are dense among all the pairs in some fiber. So we have:

Corollary 6.26. Let (S, G, S^*) be a classical Ellis semigroup with G a group. Assume that $\Phi : S \times X \to X$, and $\Psi : S \times Y \to Y$ are classical actions and that $\pi : X \to Y$ a continuous action map. If X is minimal and π is a proximal, open map, then the restriction of Φ^2 to R_π is S^* transitive. If, in addition, G is separable then the recurrent points are dense in R_π.

Corollary 6.27. Let (S, G, S^*) be a classical Ellis semigroup with G a group. Assume that $\Phi : S \times X \to X$, and $\Psi : S \times Y \to Y$ are classical actions and that $\pi : X \to Y$ a continuous action map. Assume that X is metrizable as well as minimal and that π is an open map. If for every positive integer n $R_\pi^n = QPROX^n(\pi)$, and a fortiori if π is a proximal map, then the restriction of Φ^2 to R_π is S^* transitive and S^* point transitive.

PROOF. By Proposition 6.21 there exists for $n = 1, 2, ...$ a dense G_δ subset $D_n \subset Y$ such that for $y \in D_n$ $PROX^n \cap X_y^n$ is dense in X_y^n. By the Baire Category Theorem there exist points y in the intersection of all the D_n's. Hence, Theorem 6.25 applies and shows that the action on R_π is S^* transitive. By Theorem 5.18(e) metrizability implies that the action is point transitive. Then by Theorem 6.3 it is S^* point transitive.

If π is a proximal map then by Proposition 6.18, $R_\pi^n \subset PROX^n$ for $n = 1, 2, ...$. □

Many of the above results have demanded that the action map π be an open map. This is a rather strong condition which we would like to eliminate as a requirement. We now prove several results which do indeed allow us to omit the hypothesis that π be open, but at the cost of weakening the conclusions. Each proposition is obtained from the original with the open condition by using a construction due to Veech often referred to as the *Veech Shadow Diagram*. It provides an irreducible lifting of an action map between minimal systems to obtain such an action map which is, in addition, open.

Theorem 6.28. Let G be a topological group with minimal topological actions $\phi : G \times X \to X$ and $\psi : G \times Y \to Y$. Assume that $\pi : X \to Y$ is a continuous action map. There exists a minimal topological action $\tilde{\psi} : G \times \tilde{Y} \to \tilde{Y}$, an action map $\rho : \tilde{Y} \to Y$, and a minimal closed invariant subset $\tilde{X}$ of the pull-back $R_{\pi,\rho}$ with the following properties. With $\tilde{\pi} : \tilde{X} \to \tilde{Y}$ and $\theta : \tilde{X} \to X$ denoting the action maps which are the restrictions of the projections from $R_{\pi,\rho}$, the maps ρ and θ are irreducible and the map $\tilde{\pi}$ is open. Moreover $\tilde{X}$ is the unique minimal subset of $R_{\pi,\rho}$:

$$\begin{array}{ccc} \tilde{X} & \xrightarrow{\tilde{\pi}} & X \\ \theta \downarrow & & \downarrow \rho \\ \tilde{Y} & \xrightarrow{\pi} & Y \end{array}$$

Now assume that, in addition, X is metrizable. The maps ρ and θ are almost one-to-one. In particular, there exists a dense G_δ subset $D_\pi \subset Y$ such that $\tilde{D}_\pi =_{def} \rho^{-1}(D_\pi)$ is a dense G_δ subset of $\tilde{Y}$ and the restriction $\rho : \tilde{D}_\pi \to D_\pi$ is a homeomorphism. Furthermore, for each pair $(\tilde{y}, y) \in \tilde{D}_\pi \times D_\pi$ with $y = \pi(\tilde{y})$, the restriction $\theta : \tilde{\pi}^{-1}(\tilde{y}) \to \pi^{-1}(y)$ is a homeomorphism.

PROOF. The concepts of minimality and action map are the usual ones for group actions. For a locally compact group G this coincides with the ones we have been using for the extension to the classical actions of $(\beta_u G, G, \beta_u^* G)$ as in Example (1). If G is not originally locally compact we can replace the original topology by the discrete one.

We will outline the construction. For the omitted arguments see Glasner (1976), Veech (1977) and the proofs for Proposition A.8 and Theorem A.3 in the Appendix of Akin (1997).

From the properties of 2^X listed on page 110 we see that a topological G action on X induces a topological G action on 2^X. If $\pi : X \to Y$ is a surjective, continuous map then we let

$$2^\pi \;=_{def}\; (\pi_*)^{-1}(i_Y(Y)) \;=\; \{A \in 2^X : \pi(A) \text{ is a singleton subset of } Y\}.$$

If π is an action map of topological G actions then 2^π is a closed, invariant subset of 2^X and $\rho_0 =_{def} (i_Y)^{-1} \circ \pi_* : 2^\pi \to Y$ is a surjective action map. Clearly, $\{\pi^{-1}(y) : y \in Y\}$ is a G invariant subset of 2^π and so is its closure which we denote $\tilde{Y}_0$. It need not be true that the restriction $\rho_0 : \tilde{Y}_0 \to Y$ is an irreducible map, but one can show that there is a unique closed $\tilde{Y} \subset \tilde{Y}_0$ which is minimal among the closed subsets of $\tilde{Y}_0$ which are mapped by ρ_0 onto Y. By uniqueness, $\tilde{Y}$ is a G invariant subset. The restriction $\rho : \tilde{Y} \to Y$ of ρ_0 is the required irreducible action map.

The pullback $R_{\pi,\rho}$ consists of pairs $(x,k) \in X \times \tilde{Y}$ where the point $\pi(x) \in Y$ is the π image of k regarded as a subset of X (it is a point of 2^X). Intersecting with the closed element relation we obtain the closed invariant subset $\tilde{X}_0 =_{def} \{(x,k) : k \in Y \text{ and } x \in k\} \subset R_{\pi,\rho}$. Letting $\theta_0 : \tilde{X}_0 \to X$ denote the restriction of the projection map to the first coordinate, it is surjective and again one can show that there exists a unique closed $\tilde{X} \subset \tilde{X}_0$ which is minimal among the closed subsets of $\tilde{X}_0$ which are mapped by θ_0 onto X. The restriction $\theta : \tilde{X} \to X$ is the second required irreducible action map.

Letting $\tilde{\pi}_0 : R_{\pi,\rho} \to \tilde{Y}$ denote the restriction of the second coordinate projection one can show that $\tilde{\pi}_0$ maps $\tilde{X}$ onto $\tilde{Y}$ and so restricts to define $\tilde{\pi} : \tilde{X} \to \tilde{Y}$. Finally, one shows that $\tilde{\pi}$ is an open map.

That π is an action map of minimal systems is used in the hidden bits where the property described by Lemma 6.14(a) is applied.

If X is metrizable, and so Y and 2^X are as well, Proposition 4.10(b) implies that the irreducible map ρ is almost one-to-one and the required dense G_δ sets can be taken to be $\tilde{D}_\pi = Inj(\rho)$ and $D_\pi = \rho(Inj(\rho))$. The map ρ is open at the points of $\tilde{D}_\pi$ and so the restriction of ρ is a homeomorphism.

Since $\tilde{X}$ is a subset of the pullback $R_{\pi,\rho}$ it is easy to see that whenever $\rho(\tilde{y}) = y$ the restriction $\theta : \tilde{\pi}^{-1}(\tilde{y}) \to \pi^{-1}(y)$ is injective. Furthermore, since $\theta : \tilde{X} \to X$ is surjective

$$(6.58) \quad \pi^{-1}(y) = \bigcup \{\theta(\tilde{\pi}^{-1}(\tilde{z})) : z \in \tilde{Y} \text{ such that } \rho(\tilde{z}) = y\}.$$

It follows that if $y \in D_\pi$, and so the point $\tilde{y}$ mapping onto y is unique, then $\theta : \tilde{\pi}^{-1}(\tilde{y}) \to \pi^{-1}(y)$ is surjective and so is a homeomorphism. $\square$

REMARK 6.8. All of these actions extend to classical actions of $(\beta_u G, G, \beta_u^* G)$ as described in the first paragraph of the proof above. In applying this construction to classical actions of a classical Ellis semigroup (S, G, S^*), we will always lift to $(\beta_u G, G, \beta_u^* G)$ by the surjective, continuous homomorphism $\gamma : \beta_u G \to S$ so that the semigroup actions automatically extend to the new parts of the Veech shadow diagram.

Theorem 6.29. Let (S, G, S^*) be a classical Ellis semigroup with G a group. Let $\Phi : S \times X \to X, \Psi : S \times Y \to Y$ be classical actions with X minimal and let $\pi : X \to Y$ be a continuous action map. Assume either that G is separable or that X is metrizable. If every recurrent point in R_π is a diagonal point, or, equivalently, if π is both proximal and semidistal, then π is an irreducible map. When X is metrizable π is an almost one-to-one map.

PROOF. The assumption that π is proximal and semidistal clearly implies that the only recurrent points of R_π lie in the diagonal 1_X. Conversely, the latter condition obviously implies that π is semidistal. Furthermore, if $z \in R_\pi$ and $u \in S$ is an idempotent then uz is a recurrent element of R_π and so lies in the diagonal. Hence, z is a proximal pair. Finally, by Proposition 4.5 (d) it is further equivalent to assume that π is $[Id(S)S]$ asymptotic.

If $S^* = S$ then every point is recurrent and so the assumption yields $R_\pi = 1_X$ and the map π is a homeomorphism which is certainly irreducible. So we may assume that S^* is a proper subset of S. By using the homomorphism $\gamma : \beta_u G \to S$ we pull the action back to $(\beta_u G, G, \beta_u^* G)$. By Lemma 1.5 the map π is proximal with respect to the $\beta_u G$ action. By Proposition 6.1(g) $\gamma(\beta_u^* G)) = S^*$ and so by Lemma 1.5 the map is semidistal with respect to $(\beta_u G, G, \beta_u^* G)$. The conclusion about π is topological and so is unaffected by the change.

Observe first that if, in addition, π is an open map, then Corollary 6.26 (when G is separable) and Corollary 6.27 (when X is metrizable) imply that the recurrent points are dense in R_π. By hypothesis all of the recurrent points lie in the closed diagonal subset $1_X \subset R_\pi$. This implies that $1_X = R_\pi$ which means that π is injective and so is a homeomorphism. For the general case, we lift to an open map by using the Veech construction.

Next notice that both proximality and semidistality are preserved by pullback. That is, if $\tilde{\Psi} : \beta_u G \times \tilde{Y} \to \tilde{Y}$ is an Ellis action and $\rho : \tilde{Y} \to Y$ is an action map then with $\tilde{\pi}_0 : R_{\pi,\rho} \to \tilde{Y}$ the second coordinate projection from the pullback, we have that $\tilde{\pi}_0$ is proximal and semidistal. To see this let $((x_1, k), (x_2, k))$ be a pair in $\tilde{\pi}$ fiber over $k \in \tilde{Y}$. Since $\pi(x_1) = \rho(k) = \pi(x_2)$ the pair (x_1, x_2) is in R_π and so proximal because π is a proximal map. There exists $p \in \beta_u G$ such that $px_1 = px_2$ and so $p(x_1, k) = p(x_2, k)$. On the other hand, if the pair $((x_1, k), (x_2, k))$ is recurrent in $R_{\tilde{\pi}}$ then by Lemma 1.6(a), $(x_1, x_2) \in R_\pi$ is recurrent. Because π is semidistal, $x_1 = x_2$ and so $(x_1, k) = (x_2, k)$. Finally, the restriction of $\tilde{\pi}_0$ to any closed invariant subset is a proximal, semidistal action map.

It follows that when we perform the Veech lifting, the map $\tilde{\pi} : \tilde{X} \to \tilde{Y}$ is proximal and semidistal as well as open. As noted above, Corollaries 6.26 and 6.27 now imply that $\tilde{\pi}$ is a homeomorphism and so the composition $\pi \circ \theta = \rho \circ \tilde{\pi}$ is irreducible because ρ is. From Proposition 4.10(d) it follows that π is irreducible.

When X is metrizable the irreducible map is almost one-to-one by Proposition 4.10(b). $\square$

Compare the following result with Proposition 2.16.

Corollary 6.30. Let Let (S, G, S^*) be a classical Ellis semigroup with G a group. Let $\Phi : S \times X \to X, \Psi : S \times Y \to Y$ be classical actions with X minimal and let $\pi : X \to Y$ be a continuous action map. Assume either that G is separable or that X is metrizable. If π is an asymptotic map, then

π is an irreducible map. If, in addition, X is metrizable then π is almost one-to-one.

PROOF. An asymptotic map is proximal and semidistal. □

Compare the following with Proposition 6.23. This *Dichotomy Theorem* is a topological version of the well known fact that for a homomorphism of ergodic measure preserving transformations either almost all fibers have a constant, finite cardinality or almost all fibers have the cardinality of the continuum (in fact, are atomless standard Lebesgue space).

Theorem 6.31. Let (S, G, S^*) be a classical Ellis semigroup with G a group. Assume that $\Phi : S \times X \to X$ and $\Psi : S \times Y \to Y$ are classical actions and that $\pi : X \to Y$ is a continuous action map. Assume that X is minimal and metrizable. Exactly one of the following is true.
  (1) For some positive integer n the set $\{y \in Y : \pi^{-1}(y)$ has cardinality $n\}$ is a dense G_δ subset of Y and every other fiber $\pi^{-1}(y)$ is either infinite or has finite cardinality greater than n.
  (2) Every fiber $\pi^{-1}(y)$ is infinite and $\{y \in Y : \pi^{-1}(y)$ is perfect $\}$ is a dense G_δ subset of Y.

PROOF. : Choose a metric d on X.

First assume that some fiber is finite and let n denote the minimum cardinality among the fibers. Define $O_{n,\epsilon}$ to be the set of points $y \in Y$ such that $\pi^{-1}(y)$ is a subset of a union of n open subsets of X each of diameter less than ϵ. By compactness each $O_{n,\epsilon}$ is open and the G_δ set O_n obtained by intersecting over positive ϵ's consists of those points of Y whose fibers have cardinality at most n. By choice of n this set is nonempty and consists of fibers with cardinality exactly n. Since the elements of G map fibers to fibers bijectively, the set O_n is G invariant and so is dense because Y is minimal. This is Case (1) of the theorem.

Now assume that all fibers are infinite and assume, in addition, that π is open. Set
$$(6.59) \qquad f(x) = \sup\{\epsilon \geq 0 : B_\epsilon(x) \cap \pi^{-1}(\pi(x)) = \{x\}\}.$$
Note that $f(x) > 0$ iff x is an isolated point of $\pi^{-1}(\pi(x))$. We will show that $f : X \to [0, \infty)$ is upper-semi-continuous, i.e. for every ϵ the set $\{x : f(x) \geq \epsilon\}$ is closed. In fact suppose $f(x_n) \geq \epsilon > 0$ with $x_n \to x$ and suppose that for some point $x' \in X$ with $x \neq x'$, $\pi(x') = \pi(x)$ we have $d(x, x') < \epsilon$. By openness of π there exists a sequence $x'_n \to x'$ with $\pi(x'_n) = \pi(x_n)$. We have $d(x_n, x'_n) \leq d(x_n, x) + d(x, x') + d(x'x'_n)$ and it follows that eventually $d(x_n, x'_n) < \epsilon$. This contradicts our assumption that $f(x_n) \geq \epsilon$ for every n.

It follows from the upper-semi continuity of f that the set X_c of continuity points of f is a dense G_δ subset of X. Suppose there is a point $x_0 \in X_c$ with $f(x_0) = \epsilon > 0$. Because x_0 is a continuity point, these exists $\delta < \epsilon/2$ such that for $x \in U = B_\delta(x_0)$ $f(x) > \epsilon/2$. Hence, $R_\pi \cap (U \times U) = 1_U$.

Let $V = G^{-1}(U \times U)$. By minimality V is a neighborhood of the diagonal $1_X \subset X \times X$ and $R_\pi \cap V = 1_X$. We conclude that every point $x \in X$ is an isolated point of the set $\pi^{-1}(\pi(x))$. By compactness we conclude that each fiber is finite and since we already ruled out this possibility we have now shown that $f(x) = 0$ for every $x \in X_c$.

We are still assuming that π is open and so we can apply the Fubini Theorem 6.21 to get the dense G_δ subset Y_c such that $\pi^{-1}(y) \cap X_c$ is dense in $\pi^{-1}(y)$ for every $y \in Y_c$. Any dense set contains all the isolated points but $\pi^{-1}(y) \cap X_c$ contains no isolate points of $\pi^{-1}(y)$. Hence, for $y \in Y_c$ the fibers have no isolated points and so are perfect.

Now we extend to general π by using the Veech construction. Let $D_\pi \subset Y$ and $\tilde{D}_\pi \subset \tilde{Y}$ be the dense G_δ sets described in the statement of Theorem 6.28.

We are assuming that the fibers of π are all infinite and we first observe that the same is true of $\tilde{\pi}$. Otherwise by Proposition 6.23 all of the fibers of $\tilde{\pi}$ would be finite. On the other hand the fibers over points of $\tilde{D}_\pi$ are mapped bijectively onto the - infinite - fibers over the corresponding points of D_π by θ.

Thus, we can apply the above results to obtain the dense G_δ set $\tilde{Y}_c$ with perfect $\tilde{\pi}$ fibers. By the Baire Category Theorem $\tilde{Y}_c \cap \tilde{D}_\pi$ is a dense G_δ subset of $\tilde{D}_\pi$ which is mapped homeomorphically onto D_π by ρ. Hence, $\rho(\tilde{Y}_c \cap \tilde{D}_\pi)$ is a dense G_δ subset of D_π and so of Y. Since θ is a homeomorphism for the $\tilde{D}_\pi$ fibers it follows the π fibers over the points of $\rho(\tilde{Y}_c \cap \tilde{D}_\pi)$ are perfect. □

For the next result we define a *Mycielski subset* of a compact metric space X to be a dense subset which is a countable union of nowhere dense Cantor subsets of X. A compact metric space admits Mycielski subsets iff it is perfect, see Blanchard et al. (2002) and Akin (2004) Section 2. The following is the Kuratowski-Mycielski Theorem, Theorem 5.10 of Akin (2004), applied to the set $Q \subset 2^X$ consisting of all $K \in 2^X$ such that $K \times K \subset P$. See also Huang and Ye (2002).

Theorem 6.32. Let X be a perfect, compact metric space and P be a dense, G_δ subset of $X \times X$ such that

(6.60) $\qquad (x, y) \in P \quad \Rightarrow \quad (y, x), (x, x) \in P.$

There exists a Mycielski subset $B \subset X$ such that $B \times B \subset P$.

Theorem 6.33. Let (S, G, S^*) be a classical Ellis semigroup with G a group. Let $\Phi : S \times X \to X, \Psi : S \times Y \to Y$ be classical actions and

$\pi : X \to Y$ be a continuous action map. Assume that X is is minimal and metrizable and that π is proximal but <u>not</u> almost one-to-one. Then every fiber $\pi^{-1}(y)$ is infinite and there exists a dense G_δ subset D of Y such that for every $y \in D$ the fiber $\pi^{-1}(y)$ is perfect and, in addition, there exists B_y a Mycielski subset of $\pi^{-1}(y)$ such that for all $x_1, x_2 \in B_y$ the pair (x_1, x_2) is both proximal and recurrent.

PROOF. : We proceed as before. First we lift to get a classical action of $(\beta_u G, G, \beta_u^* G)$ and then we use the Veech shadow diagram of Theorem 6.28.

As observed in the proof of Theorem 6.29, π proximal implies that $\tilde\pi$ is proximal. Hence, $\tilde\pi$ is proximal and open and so we can apply Proposition 6.23 to $\tilde\pi$. If any fiber is finite $\tilde\pi$ is an $n-$to-one distal map. The only way it can also be proximal is if $n = 1$. In that case, $\tilde\pi$ is a homeomorphism and, as shown in the proof of Theorem 6.29, π is almost one-to-one. Hence, every fiber is infinite and by Theorem 6.31 there exists a dense G_δ set $\tilde D_1 \subset \tilde Y$ over which the fibers are perfect.

Corollary 6.27 also applies to $\tilde\pi$ and so the product action on $R_{\tilde\pi}$ is point transitive with $TRANS = TRANS_{\beta^*}$ a dense G_δ subset of $R_{\tilde\pi}$ by Theorem 5.18(e). Since $TRANS = TRANS_{\beta^*}$ each transitive point is recurrent and so is recurrent for the original (S, G, S^*) action as well. If we let $P = TRANS \cup 1_{\tilde X}$ then P is a dense G_δ subset of $R_{\tilde\pi}$. By Lemma 6.24 the map $\lambda_{\tilde\pi} : R_{\tilde\pi} \to \tilde Y$ is open and so by the Fubini Theorem 6.21 there is a dense G_δ subset $\tilde D_2$ of $\tilde Y$ such that for each $y \in \tilde D_2$ the subset $\tilde\pi^{-1}(y) \times \tilde\pi^{-1}(y)$ of $R_{\tilde\pi}$ intersects P in a dense subset.

Since we are in the metrizable case, the irreducible map $\rho : \tilde Y \to Y$ is almost one-to-one. Let $\tilde D_\pi$ denote $Inj(\rho)$ and let D_π denote its image $\pi(\tilde D_\pi)$. These are dense G_δ subsets of $\tilde Y$ and Y respectively. Recall that if $y \in D_\pi$, and so the point $\tilde y$ mapping onto y is unique, then the map $\rho : \tilde\pi^{-1}(\tilde y) \to \pi^{-1}(y)$ is a homeomorphism.

Let $\tilde D = \tilde D_1 \cap \tilde D_2 \cap \tilde D_\pi$ which is a dense G_δ subset of $\tilde Y$ by the Baire Category Theorem. Let $D = \rho(\tilde D)$ which is a dense G_δ subset of D_π since $\rho : \tilde D_\pi \to D_\pi$ is a homeomorphism. Since D_π is itself a dense G_δ subset of Y, so is D.

Now we are ready to reach our conclusions.

Since every fiber of $\tilde\pi$ is infinite and since the surjection θ is injective on fibers it follows that every fiber of π is infinite.

Now assume $y \in D$. Since $y \in D_\pi$ there is a unique point $\tilde y$ with $\rho(\tilde y) = y$ and $\tilde y \in \tilde D_1 \cap \tilde D_2 \cap \tilde D_\pi$. Since $\tilde y \in \tilde D_1$ the fiber $\tilde\pi^{-1}(\tilde y)$ is perfect and the homeomorphism θ shows that $\pi^{-1}(y)$ is perfect. Since $\tilde y \in \tilde D_2$ it follows that $P_y =_{def} \theta \times \theta(P \cap [\tilde\pi^{-1}(\tilde y) \times \tilde\pi^{-1}(\tilde y)])$ is a dense G_δ set of recurrent points in $\pi^{-1}(y) \times \pi^{-1}(y)$. Clearly, P_y satisfies condition (6.61). By Theorem 6.32 there exists a Mycielski subset B_y of $\pi^{-1}(y)$ such that $B_y \times B_y \subset P_y$ and so all the pairs drawn from B_y are recurrent. All pairs in the fiber $\pi^{-1}(y)$ are proximal. Thus, B_y satisfies the required conditions. □

REMARK 6.9. In the Introduction we called a set B *strongly scrambled* when every pair drawn from B is both proximal and recurrent. A system is called *strongly Li-Yorke chaotic* when it admits an uncountable strongly scrambled set. The above result shows that if a metric system admits a proximal map which is not almost one-to-one then the system is strongly Li-Yorke chaotic. Indeed almost every fiber contains such a strongly scrambled set.

We conclude with a pair of unrelated results.

In the group case some of the results for $[Min(S)]$ semidistality are in fact well known for so-called *PI systems*. These are proximal factors of minimal systems which can be obtained from a trivial system as a, possibly transfinite, sequence of lifts by proximal and distal action maps. The connection is given by the following characterization due to Bronstein and van der Woude.

Theorem 6.34. Let (S, G, S^*) be a classical Ellis semigroup with G a group. A minimal classical action $\Phi : S \times X$ is PI iff it is $[Min(S)]$ semidistal.

PROOF. The characterization, due to Bronstein (1977) in the metric case and to van der Woude (1985) in general, (see also Auslander (1988), Chapter XIV, Theorem 31 and de Vries (1993) Chapter VI) says that a minimal dynamical system is PI iff every point transitive subsystem of the product $X \times X$ which has a dense set of minimal points is in fact minimal. This exactly says that every $[Min(S)]$ recurrent point in $X \times X$ is minimal. By Theorem 4.6(b) this holds iff Φ is $[Min(S)]$ semidistal. □

Finally, we describe the promised converse to Proposition 2.17(c).

Theorem 6.35. Let (S, G, S^*) be a classical Ellis semigroup with G a group and let $\Phi : S \times X \to X$ be a classical action. A point $x \in X$ is a distal point iff it is a product recurrent point.

PROOF. We forget the topology on G and lift the action to $(\beta G, G, \beta^* G)$ as in the proof of Theorem 6.18. Then apply Corollary 5.36 of Ellis et al. (2000) which extends the abelian group results of Auslander and Furstenberg (1994). □

CHAPTER 7

Classical Actions: The Abelian Case

In this section we return to the case where G need not be a group in the classical Ellis semigroup (S, G, S^*) but we assume that G is abelian.

Definition 7.1. Let (S, G, S^*) be a classical Ellis semigroup with G abelian. We say that Let (S, G, S^*) satisfies the *Intersection Condition* if

- S^* is the complement of G in S, i.e. $S^* = S \setminus G$.
- For every $g \in G$ the image set gS is a neighborhood of S^* in S.
- S^* is the intersection of these images, i.e.

(7.1) $$S^* = \bigcap_{g \in G} gS.$$

Proposition 7.2. Let (S, G, S^*) be a classical Ellis semigroup with G abelian.

(a) If $g \in G$ then

(7.2) $$gp = pg \quad \text{for all} \quad p \in S$$
$$\text{i.e.} \quad M^g = M_g \quad \text{on} \quad S.$$

(b) $[Min(S)] \subset S^*$ is the S closure of the ideal $Min(S)$ of minimal points in S. $[Id(S)S]$ is the S closure of the set $Id(S)S = RECUR(S)$ of recurrent points in S.

(c) If (S, G, S^*) satisfies the Intersection Condition then (S, G, S^*) satisfies the Surjection Condition. If, in addition, G is separable then S^* is a G_δ subset of S.

(d) (S, G, S^*) satisfies the Intersection Condition iff $\{gS : g \in S\}$ is a basis for the filter of neighborhoods in S, or, equivalently, iff

(7.3) $$k\mathcal{B} = [[\{gG : g \in G\}]]$$

(e) If $\Phi : S \times X \to X$ is a classical action the set $Min(X)$ of minimal points is invariant and its closure is the min-center $[Min(X)]$. The set $RECUR = Id(S)X$ is G invariant and its closure is the center $[Id(S)X]$.

(f) If $\Phi : S \times X \to X$ is a classical action and $g \in G$ then $\Phi^g : X \to X$ is a continuous action map.

PROOF. (a): By assumption, equation (7.2) holds for all $p \in G$. Because G is dense and the left and right translations by g are continuous, it holds for all $p \in S$.

(b), (e): The results for $[Min(S)]$ and $[Min(X)]$ are a restatement of Corollary 5.7. Now $px = x$ and (7.2) imply that $pgx = gpx = gx$ for all $g \in G$. That is, p fixes each point of the G orbit Gx which is dense in Sx. Thus,

$$(7.4) \qquad Iso_x \;\subset\; Iso_{gx},$$

for all $x \in X, g \in G$ and the set $RECUR = Id(S)X$ is G invariant. So its closure is invariant because the action is densely continuous.

(c),(d): Now for F any finite subset of G let g_F be the product of the elements of F. G is abelian, so the order doesn't matter. Observe that

$$(7.5) \qquad \begin{aligned} F_1 \subset F_2 &\implies g_{F_2}S \subset g_{F_1}S, \\ \text{and so for all} \quad & F_1, F_2 \in Fin(G) \\ g_{F_1 \cup F_2}S = g_{F_1}g_{F_2}S &\subset g_{F_1}S \cap g_{F_2}S. \end{aligned}$$

If A is any neighborhood of S^* it follows from compactness and (7.1) that some $g_F S$ is contained in A. Since these are all neighborhoods of S^*, they form a neighborhood base. This is equivalent to saying that the filter of neighborhoods is generated as a family by these sets. Because $S^* = S \setminus G$ is an ideal:

$$(7.6) \qquad gS \cap G \;=\; gG \quad \text{for all} \quad g \in G.$$

So, we obtain (7.3).

Now assume G is separable and choose G_0 a countable dense subsemigroup of G. If A is an open neighborhood of S^* and $gS \subset A$ then there is a neighborhood $B \subset G$ of g such that $BS \subset A$ by compactness of S and continuity of the G action. There exists $h \in B \cap G_0$ and so $hS \subset A$. Thus, we can obtain a countable neighborhood base for S^* by restricting F to $Fin(G_0)$. It follows that S^* is G_δ.

Now for each $q \in S^*$ and $g \in G$, let $K(q, F) = \{p \in g_F S = Sg_F : gp = pg = q\}$. This is a closed subset of S and since $S^* \subset gg_F S$ it is nonempty. Furthermore, (7.5) implies that as F varies over the finite subsets of G the collection of $K(q, F)$ satisfies the finite intersection property. There exists p a point of the intersection. By assumption (7.1), $p \in S^*$ with $gp = q$. Thus, (S, G, S^*) satisfies the Surjection Property.

(f): By (7.2), $pgx = gpx$ for all $p \in S$ and $x \in X$. □

Proposition 7.3. Let (S, G, S^*) be a classical Ellis semigroup with G abelian and let $\Phi : S \times X \to X$ be a classical Ellis action.

7. CLASSICAL ACTIONS: THE ABELIAN CASE

(a) The prolongations $\mathcal{N}\Phi$ and $\mathcal{N}^*\Phi$ are closed, invariant subsets of $X \times X$. For all $x \in X$ and $p \in S$

(7.7)
$$p\mathcal{N}\Phi(x) \subset \mathcal{N}\Phi(px) \cap \mathcal{N}\Phi(x).$$
$$p\mathcal{N}^*\Phi(x) \subset \mathcal{N}^*\Phi(px) \cap \mathcal{N}^*\Phi(x).$$

(b) If (S, G, S^*) satisfies the Surjection Condition then $\mathcal{N}\Phi$ and $\mathcal{N}^*\Phi$ are $S \times S$ invariant. That is, for all $p, q \in S$

(7.8)
$$(\Phi^p \times \Phi^q)(\mathcal{N}\Phi) \subset \mathcal{N}\Phi. \quad \text{and} \quad (\Phi^p \times \Phi^q)(\mathcal{N}^*\Phi) \subset \mathcal{N}^*\Phi.$$

Furthermore, if either $q \in S^*$ then

(7.9)
$$(\Phi^p \times \Phi^q)(\mathcal{N}\Phi) \subset \mathcal{N}^*\Phi.$$

For each $x \in X$ the subsystem obtained by restricting to S^*x has a surjective G action.

(c) If (S, G, S^*) satisfies the Intersection Condition then for all $x \in X$ and $g \in G$

(7.10)
$$g\mathcal{N}^*\Phi(gx) = \mathcal{N}^*\Phi(x) \subset \mathcal{N}^*\Phi(gx),$$

and so the subsystem obtained by restricting to $\mathcal{N}^*\Phi(x)$ has a surjective G action.

(d) The nonwandering set $|\mathcal{N}^*\Phi|$ is a closed, invariant subset of X. If x is a nonwandering point then $\mathcal{N}\Phi(x) = \mathcal{N}^*\Phi(x)$. If (S, G, S^*) satisfies the Surjection Condition then

(7.11)
$$S^*x \subset |\mathcal{N}^*\Phi| \quad \text{for all} \quad x \in X.$$

(e) The regional proximality relation $QPROX(\Phi)$ is a closed, invariant subset of $X \times X$.

PROOF. (a), (b): Observe that for $U, V \subset X$ and for $g \in G$

(7.12)
$$N^\Phi((\Phi^g)^{-1}(U), (\Phi^g)^{-1}(V)) = N^\Phi(\Phi^g((\Phi^g)^{-1}(U)), V)$$
$$\subset N^\Phi(U, V)$$

with equality when Φ^g is surjective. It follows that $(x, y) \in \mathcal{N}\Phi$ implies $(px, py) \in \mathcal{N}\Phi$ for all $p \in G$ and so for all $p \in S$ since $\mathcal{N}\Phi$ is closed. Furthermore, Proposition 5.12 implies that $(x, py) \in \mathcal{N}\Phi$ for all $p \in S$. Similarly, for $\mathcal{N}^*\Phi$. The formulae of (7.7) just restate these results.

Now assume that the Surjection Condition holds and let $g \in G, q \in S$ and $(x, y) \in \mathcal{N}\Phi$. By the previous remarks, $(gx, gpy) \in \mathcal{N}\Phi$ for all $p \in S$. By the Surjection Condition there exists p such that $gp = q$. Hence, $(gx, qy) \in \mathcal{N}\Phi$. Again closure implies that $(px, qy) \in \mathcal{N}\Phi$ for all $p, q \in S$. Furthermore, we use $\mathcal{N}\Phi(x) = Sx \cup \mathcal{N}^*\Phi(x)$, i.e. (5.44), to show that $q \in S^*$ implies $(x, y) \in \mathcal{N}^*\Phi$ exactly as in the proof of Proposition 6.2 (a). The argument for $S \times S$ invariance of $\mathcal{N}^*\Phi$ is the same as that for $\mathcal{N}\Phi$.

Since the Surjection Condition implies $S^*x = gS^*x$, the subsystem on S^*x has a surjective G action.

(d): Now assume the Intersection Condition. Let $y \in \mathcal{N}^*\Phi(x)$ and let $\mathcal{T}_x$ and $\mathcal{T}_y$ be the filters of neighborhoods for x and y, respectively. For every, $(U, V, L) \in \mathcal{T}_x \times \mathcal{T}_y \times k\mathcal{B}$, the hitting time set $N(U, V)$ lies in $\mathcal{B}$. Since $gG = gS \cap G$ lies in the filter $k\mathcal{B}$ it follows that the intersection $N(U,V) \cap L \cap gG$ is nonempty and so we can choose $h(U, V, L) \in G$ and $x(U, V, L) \in U$ such that $gh(U, V, L) \in N(U, V) \cap L$ and $gh(U, V, L)x(U, V, L) \in V$. The net $\{x(U, V, L)\}$ indexed by the directed set $\mathcal{T}_x \times \mathcal{T}_y \times k\mathcal{B}$ converges to x and we can choose a subnet of the $h(U, V, L)x(U, V, L)$ which converges to a point $z \in X$. Clearly, $gz = \text{Lim } gh(U,V,L)x(U,V,L) = y$ and since L varies over $k\mathcal{B}$, we have $z \in \mathcal{N}^*\Phi(x)$. Thus, $\mathcal{N}^*\Phi(x) \subset g\mathcal{N}^*\Phi(x)$. (7.10) then follows from (7.7).

(e): If $(x, x) \in \mathcal{N}^*\Phi$ then by part (a) $(px, px) \in \mathcal{N}^*\Phi$. If the Surjection condition holds then by (7.9) $(px, px) \in \mathcal{N}^*\Phi$ for all $p \in S^*$ and all $x \in X$ because (x, x) is always in $\mathcal{N}\Phi$ since S is a monoid. If $x \in \mathcal{N}^*\Phi(x)$ then $Sx \subset \mathcal{N}^*\Phi(x)$ and so $\mathcal{N}^*\Phi(x) = \mathcal{N}\Phi(x)$ by (5.44).

(f): If $(z, z) \in \mathcal{N}(\Phi \times \Phi)(x, y)$ then by part (a) $(pz, pz) \in \mathcal{N}(\Phi \times \Phi)(px, py)$. □

REMARK 7.1. Since the nonwandering set is invariant we can use the inductive construction described in the group case to close down to the center of the system when G is separable (see Remark (b) after Proposition 6.2).

Recall that, for any Ellis action $\Phi : S \times X \to X$, $\Phi^\# : S \to X^X$ is a continuous homomorphism with image the enveloping semigroup. We denote by Iso_X the kernel of this homomorphism, so that
(7.13)
$$Iso_X \; =_{def} \; \{p \in S : px = x \quad \text{for all} \quad x \in X\} \; = \; \bigcap_{x \in X} Iso_x.$$

As the intersection of closed co-ideals, Iso_X is a closed co-ideal when it is nonempty. Of course, for any classical action, or, more generally, for any monoid action, the identity e lies in Iso_X.

Theorem 7.4. Let (S, G, S^*) be a classical Ellis semigroup with G abelian and let $\Phi : S \times X \to X$ be a classical Ellis action.

(a) Assume that Φ is point transitive with surjective G action. $TRANS$ is a dense, G invariant subset of X and Φ is transitive.

Φ is S^* transitive iff it is S^* point transitive. If (S, G, S^*) satisfies the Surjection Condition then Φ is S^* transitive and S^* point transitive.

(b) The system Φ satisfies exactly one of the following conditions.

Case (0) Φ is neither transitive nor point transitive.

Case (1) Φ is S^* transitive and S^* point transitive. If (S, G, S^*) satisfies the Surjection Condition then, in this case, $TRANS = TRANS_{S^*} \subset RECUR$ is a dense, G invariant subset and the system Φ has a surjective G action.

Case (2) Φ is S^* transitive but not point transitive, i.e. $TRANS = \emptyset$. If (S, G, S^*) satisfies the Surjection Condition with G separable then the system Φ has a surjective G action and the set $RECUR(\Phi)$ of recurrent points is dense in X. When X is metrizable this case does not occur.

Case (3) Φ is point transitive but not S^* point transitive or S^* transitive. In this case, there is a proper closed invariant set $X^* \subset X$ such that $S^* x \subset X^*$ for all $x \in X$ with equality if $x \in TRANS$. Furthermore, for all $x \in TRANS$ $X = Gx \cup X^*$. In particular, $TRANS \cap X^* = \emptyset$. The center $[RECUR(\Phi)]$ is contained in X^*.

Case (4) Φ is S^* point transitive but not transitive. In this case $TRANS$ is not a dense subset of X. If (S, G, S^*) satisfies the Surjection Condition then this case does not occur.

(c) Assume that Φ is point transitive but not S^* point transitive or S^* transitive, i.e. Case (3) above. For every $x \in TRANS$, $Iso_x = Iso_X$ and this compact co-ideal of S is contained in G. Define $H(\Phi)$ to be

$$\{p \in S : \Phi^p(TRANS) = TRANS\}$$
(7.14) $= \{p \in S : \Phi^p(TRANS) \cap TRANS \neq \emptyset\}$
$= \{g \in G : Iso_X \cap gG \neq \emptyset\}.$

$H(\Phi)$ is a co-ideal of S with $Iso_X \subset H(\Phi) \subset G$. For any $x \in TRANS$, $H(\Phi)x = TRANS$. If, in addition, (S, G, S^*) satisfies the Intersection Condition then $H(\Phi)$ is a compact subsemigroup of S disjoint from S^* and $TRANS$ is a compact subset of X disjoint from X^*. Furthermore, the G action of Φ is not surjective, Φ is not transitive and the transitive points are not dense in X.

(d) If S^* is a proper subset of S then the translation action $M : S \times S \to S$ is point transitive but not S^* point transitive. The identity e is a transitive point and $Iso_e = Iso_S = \{e\}$. $TRANS(M)$ is the group of units of the monoid G, i.e.

(7.15) $\quad TRANS(M) \quad = \quad \{g \in G : gh = e \quad \text{for some} \quad h \in G\}.$

If (S, G, S^*) satisfies the Intersection Condition then the group of units of G, which is $TRANS(M) = H(M)$, is a compact subgroup of G disjoint from S^*.

PROOF. (a): Let $x \in TRANS$. If $g \in G$ then $X = Sx$ implies $X = \Phi^g(X) = gSx = Sgx$ because the action is surjective. Thus, $gx \in TRANS$.

Density and transitivity follows from Theorem 5.18(c) which also says that S^* transitivity follows from S^* point transitivity. On the other hand, Proposition 5.19 implies that point transitivity plus S^* transitivity implies S^* point transitivity. Notice that (7.2) and (5.2) imply that the multiplication $M : S \times G \to S$ is continuous.

For the Intersection Condition result, see below.

(b): The five cases come from Theorem 5.18(g) with the following refinements. First, Proposition 5.19 implies that in Case (3) the system is not S^* transitive. Since $S^*X = X^*$ the set of recurrent points is contained in X^* as is its closure. In Case (2) with G separable we can express Φ as the surjective inverse limit of metrizable systems. By Theorem 5.18(e) the factors are point transitive and S^* transitive. By the previous remarks they are not of Case (3) and so they are of Case (1) which has dense recurrent points. Having dense recurrent points is a residual property and so Φ has dense recurrent points. In Case (4) the system cannot be transitive by Proposition 5.19 again.

(c): If $x \in TRANS$ then it is not recurrent and so $Iso_x \cap S^* = \emptyset$. Hence, $g \in Iso_x$ implies $g \in G$. So for every $p \in S$, we have $gpx = pgx = px$. Since x is a transitive point, $Sx = X$, i.e. $g \in Iso_X$.

Now assume $pTRANS \cap TRANS \neq \emptyset$. For some $x \in TRANS$, $px \in TRANS$. Since S^*x is a proper subset of X, we must have $p \in S \setminus S^*$. Since px is a transitive point, we have $gpx = x$ for some $g \in S \setminus S^*$. In particular, $gp = pg \in Iso_x = Iso_X$.

Any other member of $TRANS$ is of the form hx for some $h \in G$. Because $TRANS$ satisfies the capturing property, $g(phx) = h(gpx) = hx \in TRANS$ implies $phx \in TRANS$. Because g maps the transitive point px to the transitive point x we similarly have $ghx \in TRANS$. Hence, $p(ghx) = hx$. Thus, $pTRANS = TRANS$. It is clear that $H(\Phi) = \{p : pTRANS = TRANS\}$ is a subsemigroup and we have just seen that it is contained in G, is equal to $\{p : pTRANS \cap TRANS \neq \emptyset\}$ and is contained in $\{g \in G : gS \cap Iso_X \neq \emptyset\}$.

If $p, qp \in H(\Phi)$ then for $x \in TRANS$, $px, qpx \in TRANS$. Since q maps the transitive point px into $TRANS$, $q \in H(\Phi)$. Hence, $H(\Phi)$ is a co-ideal. It is obvious that Iso_X is contained in $H(\Phi)$. Hence, if $gS \cap H(\Phi) \neq \emptyset$ then for some $p \in S$, $gpx = pgx = x$. By the capturing property for $TRANS$, $gx \in TRANS$ and so $g \in H(\Phi)$. This completes the proof of (7.14). In fact, since G is abelian and $H(\Phi)$ is a co-ideal, we have, for all $g \in G$:

(7.16) $\qquad gS \cap H(\Phi) \neq \emptyset \quad \Longrightarrow \quad g \in H(\Phi).$

If $x, y \in TRANS$ then there exists $p \in S$ such that $px = y$ and any such p lies in $H(\Phi)$. Hence,

(7.17) $\qquad TRANS \;=\; H(\Phi)x \qquad$ for all $\; x \in TRANS$

Now assume the Intersection Condition. We use it to show that the S closure of $H(\Phi)$ is contained in G and so that $H(\Phi)$ is bounded in G. Suppose not. Then $H(\Phi)$ meets every neighborhood of S^* and so meets every gS by the Intersection Condition. From (7.16) we would then have $H(\Phi) = G$. From (7.14) we would have $gS \cap Iso_X \neq \emptyset$ for all $g \in G$ and so Iso_X would be unbounded. But $Iso_X = Iso_x$ is closed in S and disjoint from S^*.

Now let $M : G \times G \to G$ be the continuous multiplication on G. Observe that the set $Q =_{def} \{(g_1, g_2) : g_1 g_2 \in Iso_X\} = M^{-1}(Iso_X)$ is closed in $G \times G$ and it is contained in $H(\Phi) \times H(\Phi)$ by (7.16). Thus, this closed set is bounded in the product and so is compact. By (7.14) the first coordinate projection of Q is $H(\Phi)$ and so $H(\Phi)$ is compact. It follows that $TRANS = \Phi_x(H(\Phi))$ is compact in X and is disjoint from X^*. In particular, $TRANS$ is not dense.

Since $TRANS = \Phi_x(H(\Phi))$ is disjoint from X^*, $H(\Phi)$ is disjoint from $(\Phi_x)^{-1}(X^*)$. For each $g \in G \setminus H(\Phi)$ choose K_g a compact neighborhood of g in $G \setminus H(\Phi)$. Let U_g be the S interior of the compact set $K_g S$. By the Intersection Condition $S^* \cup \{g\}$ is contained in the open set U_g. It follows that $\{U_g : g \in G \setminus H(\Phi)\}$ is an open cover of the compact set $(\Phi_x)^{-1}(X^*)$. Let F be a finite subset of $G \setminus H(\Phi)$ such that $(\Phi_x)^{-1}(X^*) \subset U_F =_{def} \bigcup_{g \in F} U_g$. By compactness there exists a compact neighborhood V of X^* such that $(\Phi_x)^{-1}(V) \subset U_F$ and so $V \subset X^F =_{def} \bigcup_{g \in F} \Phi_x(K_g S) = \bigcup_{g \in F} K_g S x$. Because G is abelian, each $K_g S x$ is a compact invariant set and hence so is X^F. Because $K_g \subset G \setminus H(\Phi)$ it follows that X^F is disjoint from $TRANS$ and so is a proper subset of X. On the other hand, it contains contains V and so has nonempty interior. It follows from Theorem 5.18(a) that Φ is not transitive.

Since $\Phi^g(X) = gSx$ is disjoint from $TRANS$ for $x \in TRANS$ and $g \in G \setminus H(\Phi)$ it follows that

(7.18) $$H(\Phi) \;=\; \{g \in G : \Phi^g(X) = X\}.$$

In particular, the action is not surjective.

(a) completed: If the Intersection Condition holds then by Proposition 7.2 (c) the Surjection Condition does and so Case (4) does not occur. If the action is surjective then as was just shown, Case (3) is excluded as well. Φ is assumed point transitive which eliminates Case (0) and Case (2). Hence, Φ is in Case (1) and is S^* transitive.

(d): Since $pe = p$, we have $pe = e$ iff $p = e$. In particular, e is recurrent iff $e \in S^*$ or, equivalently, when $S^* = S$.

If $g \in TRANS$ then $gh = e$ for some $h \in S$ and since $e \notin S^*$, $g, h \notin S^*$. Hence, g is an invertible element of G. Conversely, $gh = e$ implies $g, h \in TRANS$ by the capturing property for $TRANS$. $H(M) = \{g : ge \in TRANS\}$ and so $H(M) = TRANS$. When the Intersection Condition Holds, this is compact by part (c). $\square$

REMARK 7.2. (a) The Intersection Condition is roughly the opposite extreme from the assumption that G is a group. $(\beta \mathbb{R}_+, \mathbb{R}_+, \beta^* \mathbb{R}_+)$ of Example (2) and $(\beta \mathbb{Z}_+, \mathbb{Z}_+, \beta^* \mathbb{Z}_+)$ of Example (4) satisfy the Intersection Condition. In these cases, the translation action is point transitive with the identity element 0 the unique transitive point.

(b) If (S, G, S^*) satisfies the Intersection Condition and we replace S^* by S then we lose the Surjection Condition and the Case (3) examples become Case (4) types for (S, G, S).

Example (7) Recall from Example (4), $(\mathbb{Z} \cup \{\infty\}, \mathbb{Z}, \{\infty\})$ the one point compactification of $\mathbb{Z}$. The translation a action on $X = \mathbb{Z} \cup \{\infty\}$ is transitive and point transitive with fixed point ∞. Now consider the one-point compactification $(\mathbb{Z} \times \mathbb{Z}_+ \cup \{\infty\}, \mathbb{Z} \times \mathbb{Z}_+, \{\infty\})$. This classical abelian Ellis semigroup satisfies the Surjection Condition and (7.1) but not the Intersection Condition. It acts on X by using the translation action for the points of $\mathbb{Z} \times \{0\} \cup \{\infty\}$ identified with $\mathbb{Z} \cup \{\infty\}$. The points $(n, m) \in \mathbb{Z} \times \mathbb{Z}_+$ with $m > 0$ act on X by mapping the entire space to ∞. This is a Case (3) action with $TRANS = \mathbb{Z} \subset X$ and so $TRANS$ is dense and the action is transitive as well as point transitive. The $\mathbb{Z} \times \mathbb{Z}_+$ action is not surjective.

Proposition 7.5. Let (S, G, S^*) be a classical Ellis semigroup with G abelian and let $\Phi : S \times X \to X$ be a classical action. Assume that the minimal points are dense in X.

(a) The G action of Φ is surjective.
(b) If Φ is transitive then it is S^* transitive. Furthermore, for every pair of opene $U, V \subset X$ there exists a minimal $M \subset X$ such that $M \cap U \neq \emptyset$ and $M \cap V \neq \emptyset$. If Φ is point transitive then it is S^* transitive and S^* point transitive.
(c) If $\Psi : S \times Y \to Y$ is a classical action and the minimal points are dense in Y, then the minimal points for the product action $\Phi \times \Psi$ are dense in the product $X \times Y$.

PROOF. (a): For each $g \in X$, $\Phi^g : X \to X$ is a continuous action map. If A is a closed invariant subset of X then gA is a closed, invariant subset of X which is contained in A. In particular, Φ^g maps each minimal subset onto itself. By continuity, $\Phi^g(X)$ is a compact set which contains the dense set $Min(X)$. So Φ^g is surjective.

(b): Given opene $U, V \subset X$, transitivity implies there exists $x \in U$ and $g \in G$ with $gx \in V$. By density of $Min(X)$ and continuity of Φ^g we can choose such an x with $x \in Min(X)$. Hence, $M = Sx$ is a minimal subset which meets both U and V. By Proposition 5.18(d) the subsystem obtained by restricting to M is S^* transitive. Hence, $N(U,V) \supset N(U \cap M, V \cap M) \in \mathcal{B}$. Consequently, Φ is S^* transitive.

If the system is point transitive then by Theorem 7.4(a) it is transitive. By the argument of the previous paragraph it is S^* transitive. So by Proposition 5.19 it is S^* point transitive.

(c): Fix a minimal idempotent $u \in S$.

Given opene subsets U and V of X and Y, respectively, there exist minimal subsets M and N which meet U and V respectively. For any $x \in M$, ux is a point of M fixed by u. Since G is abelian, Gux is a set of points fixed by u. Since M is minimal, Gux is dense in M. Thus, there exists a point $x \in U$ which is fixed by u. Similarly, there is a point $y \in V$ which is fixed by u. Hence, the pair $(x,y) \in U \times V$ is fixed by u and so is minimal by Proposition 2.4(b). □

Theorem 7.6. Let (S, G, S^*) be a separable, abelian classical Ellis semigroup and let $\Phi : S \times X \to X$ and $\Psi : S \times Y \to Y$ be classical actions. Assume that the minimal points are dense in X and that Ψ is a minimal, irreducible lift of a $[Min(S)]$ semidistal system. If Φ is weakly disjoint from Ψ then it is disjoint from Ψ. That is, the projection map $\pi_1 : X \times Y \to X$ is a minimal map.

PROOF. Proceed exactly as in Theorem 5.26. We need Proposition 7.5(c) to get that the minimal points are dense in $X \times Y$. Then use Theorem 4.9 in place of Theorem 2.14. □

Corollary 7.7. Assume that (S, G, S^*) is a separable, abelian classical Ellis semigroup. For a classical action $\Phi : S \times X \to X$ let P be the property that X is minimal and Φ is disjoint from all scattering systems which have dense minimal points. P is a residual property and every $[Min(S)]$ semidistal system satisfies P. If $\Psi : S \times Y \to Y$ is a classical action with Y minimal and metric and there exists a continuous, $[Min(S)]$ semidistal action map $\pi : Y \to X$ with $\Phi : S \times X \to X$ satisfying P, then Ψ satisfies P.

PROOF. Proceed as in Corollary 5.27 using Theorem 7.6 in place of Theorem 5.26. Again use Proposition 7.5(c) and use Theorem 4.9 in place of Theorem 2.14. □

We now consider weak mixing actions in the abelian case.

Lemma 7.8. Let (S, G, S^*) be a classical Ellis semigroup with G abelian. If Φ is a weak mixing classical action, then it is S^* transitive. If, in addition, (S, G, S^*) satisfies the Surjection Condition and G is separable then Φ has a surjective G action.

PROOF. Assume that Φ^2 is transitive but that Φ is not S^* transitive. In the classification of Theorem 7.4(b), Φ^2 is neither of Case (0) nor Case (4). Since the factor Φ is not S^* transitive, Φ^2 cannot be. Hence, Φ^2 must be of Case (3). The factor Φ is transitive and so is not of Case (0) or Case (4). Since it is assumed not S^* transitive, it must be Case (3) as well. From this we will derive a contradiction.

Since Φ^2 is point transitive it admits a transitive point (x, y). The points x and y are transitive points for the Case (3) action Φ. By Theorem 7.4(c), there exists $g \in H(\Phi) \subset G$ such that $gx = y$. By (7.2) $gpx = py$ for all $p \in S$. That is, the orbit $S(x, y)$ is contained in Φ^g identified with its graph as a subset of $X \times X$. Because (x, y) is a transitive point, $X \times X = \overline{S(x, y)} \subset \Phi^g$. Since Φ^g is a function, (x, y) is the unique point of the graph which projects to x. This is a contradiction because a Case (3) action is nontrivial. Thus, we have proved that a weak mixing action is S^* transitive, i.e. of Case (1) or (2). If, in addition, (S, G, S^*) satisfies the Surjection Condition and G is separable then Φ has a surjective G action in these two cases by Theorem 7.4(b). $\square$

The original Furstenberg Intersection Lemma is the abelian version. Results are easier because we have for $A, B \subset X$

$$(7.19) \quad \begin{aligned} N^\Phi(A, (\Phi^g)^{-1}(B)) &= N^\Phi(\Phi^g(A), B). \\ N^\Phi((\Phi^g)^{-1}(A), (\Phi^g)^{-1}(B)) &\subset N^\Phi(A, B). \end{aligned}$$

with equality in the latter case when the G action is surjective.

Lemma 7.9. Let (S, G, S^*) be a classical Ellis semigroup with G abelian. Let $\Phi : S \times X \to X$ be a classical action with a surjective G action. Assume that *either*: for all U, V opene subsets of X,

$$(7.20) \quad N^\Phi(U, V) \cap N^\Phi(U, U) \neq \emptyset.$$

or: for all U, V opene subsets of X,

$$(7.21) \quad N^\Phi(U, V) \cap N^\Phi(V, V) \neq \emptyset.$$

Then Φ is weak mixing.

PROOF. Since $N(U, V)$ is nonempty for opene U, V it follows that Φ is transitive.

Assume first that (7.20) holds.

7. CLASSICAL ACTIONS: THE ABELIAN CASE

Because Φ is transitive, there exist $g_1, g_2 \in G$ such that $U = U_1 \cap (\Phi^{g_1})^{-1}(V_1) \cap (\Phi^{g_2})^{-1}(U_2)$ is opene. Because the G action is surjective $V = (\Phi^{g_1})^{-1}(\Phi^{g_2})^{-1}(V_2)$ is opene. By (7.19) we have

$$N(U,V) \cap N(U,U) \subset$$
$$(7.22) \quad N((\Phi^{g_1})^{-1}(V_1), (\Phi^{g_1})^{-1}(\Phi^{g_2})^{-1}(V_2)) \cap N(U_1, (\Phi^{g_2})^{-1}(U_2))$$
$$\subset N(V_1, g_2^{-1}V_2) \cap N(U_1, g_2^{-1}U_2)$$

and so these are not empty by (7.20). If g_3 is in the latter intersection then $g_4 = g_2 g_3 \in N(V_1, V_2) \cap N(U_1, U_2)$. Thus, Φ is weak mixing.

Assume now that (7.21) holds.

There exist $g_1, g_2 \in G$ such that $V = V_1 \cap (\Phi^{g_2})^{-1}(V_2) \cap (\Phi^{g_1})^{-1}(\Phi^{g_2})^{-1}(U_2)$ is opene. Because the G action is surjective $U = (\Phi^{g_1})^{-1}(U_1)$ is opene. From (7.19)

$$N(U,V) \cap N(V,V) \subset$$
$$(7.23) \quad N((\Phi^{g_1})^{-1}(U_1), (\Phi^{g_1})^{-1}(\Phi^{g_2})^{-1}(U_2)) \cap N(V_1, (\Phi^{g_1})^{-1}(V_2))$$
$$\subset N(U_1, g_2^{-1}U_2) \cap N(V_1, g_2^{-1}V_2)$$

Finish as before. □

Theorem 7.10. Let (S, G, S^*) be a classical Ellis semigroup with G abelian. If $\Phi : S \times X \to X$ is a weak mixing classical action, then for all U_1, V_1, U_2, V_2 opene subsets of X, there exist U_3, V_3 opene subsets of X such that

$$(7.24) \qquad N^\Phi(U_3, V_3) \subset N^\Phi(U_1, V_1) \cap N^\Phi(U_2, V_2).$$

PROOF. By assumption of weak mixing, $N(U_1, U_2) \cap N(V_1, V_2) \neq \emptyset$. So we can choose $g \in G$ such that $U_3 = U_1 \cap (\Phi^g)^{-1}(U_2)$ and $V_3 = V_1 \cap (\Phi^g)^{-1}(V_2)$ are opene. Then (7.24) follows from (7.19). □

Recall the families from (6.30)

$$(7.25) \quad \begin{aligned} \mathcal{T}^\Phi &=_{def} [[\{N^\Phi(U,V) : \text{for opene } U, V \subset G \}]]. \\ \overline{\mathcal{T}}^\Phi &=_{def} \mathcal{T}^\Phi \cdot k\mathcal{B}. \end{aligned}$$

Theorem 7.11. Let (S, G, S^*) be a classical Ellis semigroup with G abelian and let $\Phi : S \times X \to X$ be a classical action.
   (a) The following conditions on Φ are equivalent.
      (1) Φ is weak mixing.
      (2) For every positive integer n the product action $\Phi^n : S \times X^n \to X^n$ is S^* transitive.

(3) For every positive integer n the product action $\Phi^n : S \times X^n \to X^n$ is transitive, i.e. Φ is weak mixing of all orders.
(4) The family $\mathcal{T}^\Phi$ is a filter, i.e. it is closed under intersection.
(5) The family $\overline{\mathcal{T}}^\Phi$ is a filter.
(6) There exists a proper filter $\mathcal{F}$ of subsets of G such that $\mathcal{T}^\Phi \subset \mathcal{F}$.
  (b) Weak mixing is a residual property stronger than S^* transitivity.
  (c) If Φ is weak mixing then for every index set I the product system $\Phi^I : S \times X^I \to X^I$ is weak mixing and so is S^* transitive. If, in addition, the G action of Φ is surjective, then Φ is scattering.
  (d) If Φ is scattering and the minimal points are dense in X then Φ is weak mixing.

PROOF. (a): (1) $\Rightarrow$ (4) follows from the Furstenberg Intersection Lemma, Theorem 7.10. The remaining equivalences are proved just as in Theorem 6.11(a). We use Lemma 7.8 in place of Lemma 6.9(b).

(b): Since transitivity is residual, this follows from Theorem 5.24(b). Note that (1) implies (2) in (a) shows that weak mixing is stronger than S^* transitivity.

(c): If $\mathcal{T}^\Phi \subset \mathcal{F}$ it is easy to check that $\mathcal{T}^{\Phi^I} \subset \mathcal{F}$ for any index set I. Hence, Φ^I is weak mixing when Φ is. Now assume that $\Psi : S \times Y \to Y$ is a minimal system with W_1, W_2 opene subsets of Y. Let U_1, U_2 be opene subsets of X. Because Y is minimal, there exists a finite $F \subset G$ such that

(7.26)
$$\bigcup_{g \in F} (\Psi^g)^{-1}(W_2) = Y$$
$$\text{and so} \quad \bigcup_{g \in F} N^\Psi(W_1, (\Psi^g)^{-1}(W_2)) = G.$$

Because Φ is weak mixing with surjective G action, the intersection of the $N^\Phi(U_1, (\Phi^g)^{-1}(U_2))$'s is nonempty. It follows that for some $g \in F$ there exists $k \in N^\Phi(U_1, (\Phi^g)^{-1}(U_2)) \cap N^\Psi(W_1, (\Psi^g)^{-1}(W_2))$ and so $gk \in N^\Phi(U_1, U_2) \cap N^\Psi(W_1, W_2)$. Hence, $\Phi \times \Psi$ is transitive.

(d) : By Proposition 5.23, Φ is weakly disjoint from itself and so Φ is weak mixing. $\square$

REMARK 7.3. (7.26) actually implies that

(7.27)
$$\bigcup_{g \in F} [N^\Phi(U_1, (\Phi^g)^{-1}(U_2)) \cap N^\Psi(W_1, (\Psi^g)^{-1}(W_2))] \supset$$
$$\bigcap_{g \in F} N^\Phi(U_1, (\Phi^g)^{-1}(U_2)),$$

and the latter set is in $\mathcal{B}$. This implies that

$$N^\Phi(U_1, (\Phi^g)^{-1}(U_2)) \cap N^\Psi(W_1, (\Psi^g)^{-1}(W_2)) \in \mathcal{B}$$

for some $g \in F$ because $\mathcal{B}$ is a filterdual. Since $gS^* \subset S^*$ the translation M^g maps elements of $\mathcal{B}$ to $\mathcal{B}$. It follows that the product action $\Phi \times \Psi$ is S^* transitive.

Theorem 7.12. Let (S, G, S^*) be a classical Ellis semigroup with G abelian and separable. Let $\Phi : S \times X \to X$ be a classical action with X minimal. The following conditions are equivalent
  (1) Φ is weak mixing.
  (2) The regional proximality relation is the entire space, i.e. $QPROX(\Phi) = X \times X$.
  (3) The proximality relation $PROX(\Phi)$ is dense in $X \times X$.

PROOF. By Proposition 7.5(a) the G action is surjective. Now proceed just as with Theorem 6.13, using Lemma 7.9 in place of Lemma 6.9(c). □

CHAPTER 8

Iterations of Continuous Maps

By a *map dynamical system* (sometimes called a *cascade*) we mean a pair (X, f) where $f : X \to X$ is a *surjective* continuous map. By iteration we obtain an action of the monoid $\mathbb{Z}_+$ which we extend to $\beta\mathbb{Z}_+$. It will be convenient to simplify the notation we will be using in this section.

Let $\mathbb{Z}_-$ denote the submonoid of nonpositive integers so that $\mathbb{Z} = \mathbb{Z}_+ \cup \mathbb{Z}_-$. Let

(8.1)
$$\begin{aligned} \beta &=_{def} \beta\mathbb{Z} \\ \beta_+ &=_{def} \beta\mathbb{Z}_+ \\ \beta_- &=_{def} \beta\mathbb{Z}_- \end{aligned}$$

Thus, $\beta_\pm$ is the closure in β of $\mathbb{Z}_\pm$. Let

(8.2)
$$\begin{aligned} \beta^* &=_{def} \beta\mathbb{Z} \setminus \mathbb{Z} \\ \beta_+^* &=_{def} \beta\mathbb{Z}_+ \setminus \mathbb{Z}_+ \\ \beta_-^* &=_{def} \beta\mathbb{Z}_- \setminus \mathbb{Z}_- \end{aligned}$$

$\beta_\pm^*$ are the two ends of β^*. That is, each is a closed ideal of β with

(8.3) $\qquad \beta^* = \beta_+^* \cup \beta_-^* \quad \text{and} \quad \beta_+^* \cap \beta_-^* = \emptyset.$

Thus, $(\beta_+, \mathbb{Z}_+, \beta^*)$ is a classical Ellis semigroup.

We regard a map dynamical system as a classical Ellis action $\Phi : \beta_+ \times X \to X$ with $f = \Phi^1$. By assumption on f Φ has a surjective $\mathbb{Z}_+$ action. Notice that $\mathbb{Z}_+$ is abelian and $(\beta_+, \mathbb{Z}_+, \beta_+^*)$ satisfies the Intersection Condition of Definition 7.1. So we can apply the results of Chapter 7 to Φ. Recall that $\Phi^\# : \beta_+ \to X^X$ is a continuous semigroup homomorphism whose image is the *enveloping semigroup of f* which we will denote by $\mathcal{E}_+(f)$ so that $\mathcal{E}_+(f) =_{def} \Phi^\#(\beta_+)$. The closed ideal $\mathcal{A}_+(f) =_{def} \Phi^\#(\beta_+^*)$ in $\mathcal{E}_+(f)$ is called the *adherence semigroup of f*. It is the enveloping semigroup of the action of $S^* = \beta_+^*$. Notice that $\{f^n\}$ is a dense sequence of continuous functions in $\mathcal{E}_+(f)$. This corresponds to the dense sequence $\mathbb{Z}_+ \subset \beta_+$. Furthermore, these functions commute with the elements of $\mathcal{E}_+(f)$ and so $f = \Phi^1 : X \to X$ is an action map by Proposition 7.2(f).

A subset A of X is f *invariant* when $f(A) \subset A$. This is the same as $\mathbb{Z}_+$ invariance with respect to Φ.

If (Y, g) is a map dynamical system and $\pi : X \to Y$ is a continuous map then it is an action map exactly when

(8.4) $\qquad \pi \circ f = g \circ \pi.$

Proximality with respect to $\beta_+, \beta_+^*, \mathcal{E}_+(f)$ and $\mathcal{A}_+(f)$ all agree. For recurrence and asymptoticity we use β_+^* or, equivalently, $\mathcal{A}_+(f)$, following the Convention of Chapter 5.

For a point $x \in X$ the orbit $\beta_+^* x = \mathcal{A}_+(f)x$ is the usual omega limit point set $\omega f(x)$ while $\beta_+ x = \mathcal{E}_+(f)x = \{f^n(x) : n \in \mathbb{Z}_+\} \cup \omega f(x)$ is the closure of the f orbit sequence of x. Following Chapter 5, we use the equivalent $\beta_+, \beta_+^*, \mathcal{E}_+(f)$ and $\mathcal{A}_+(f)$ concepts of minimal action map, minimal subset, point minimal space for the system (X, f). For the concepts of recurrent point, orbit minimality, and recurrent point minimality we use $\beta^* \mathbb{Z}_+$ or equivalently $\mathcal{A}_+(f)$.

As it is a classical action, and so is a densely continuous action, we have for Φ that a point $x \in X$ is $[Min(\beta_+)]$ recurrent iff it is recurrent, i.e. $x \in \omega f(x)$, and, in addition, the minimal points are dense in $\omega f(x)$.

We have assumed that f is surjective. By Theorem 7.4(a) any point transitive system is $\beta^* \mathbb{Z}_+$ point transitive, or equivalently $\mathcal{A}(f)$ point transitive. It follows that

(8.5) $\quad TRANS \quad =_{def} \quad \{x \in X : \omega f(x) = X\}.$

In fact, since we have the Intersection Condition and f is surjective, Cases (3) and (4) of Theorem 7.4(b) do not apply for a map dynamical system. By Theorem 7.4(b), (X, f) is transitive iff it is β_+^* transitive, or equivalently $\mathcal{A}_+(f)$ transitive. That is for all opene $U, V \subset X$

(8.6) $\quad N^f(U, V) \quad = \quad \{n \in \mathbb{Z}_+ : U \cap f^{-n}(V) \neq \emptyset\} \quad \in \quad \mathcal{B}$

where $\mathcal{B}$ is the family of infinite subsets of $\mathbb{Z}_+$ with dual $k\mathcal{B}$ the filter of cofinite subsets of $\mathbb{Z}_+$. By Proposition 7.4(a) the system (X, f) is transitive if it is point transitive, while we the converse is necessarily true only when X is metrizable. We will write $\mathcal{N}f$ and $\mathcal{N}^*f$ for the prolongation relations associated with the map dynamical system (X, f) so that $(x, y) \in \mathcal{N}f$ (or $(x, y) \in \mathcal{N}^* f$) iff $N^f(U, V) \neq \emptyset$ (resp. $N^f(U, V) \in \mathcal{B}$) for all neighborhoods U, V of x and y.

We call a map dynamical system *reversible* when f is injective and so is a homeomorphism on X. We call the system (X, f^{-1}) and its associated $(\beta_+, \mathbb{Z}_+, \beta_+^*)$ action the *reverse system* for (X, f). For a reversible system we will call the classical Ellis action $\Phi_\pm : \beta \times X \to X$ of $(\beta, \mathbb{Z}, \beta^*)$ defined by $f = \Phi_\pm^1$ and so $f^{-1} = \Phi_\pm^{-1}$ the associated *group system*. To it the results of Chapter 6 apply. We will write $\mathcal{N}\Phi_\pm$ and $\mathcal{N}^*\Phi_\pm$ for the prolongation relations associated with the group system.

For any map system (X, f) a subset $A \subset X$ is $\mathbb{Z}_+$ minus invariant in the sense of (5.14) when $fx \in A$ implies $x \in A$. When the system is reversible this says exactly that A is f^{-1} invariant, i.e. $f^{-1}(A) \subset A$. For a reversible system, we will say that A is f $\pm$invariant when it is both f and f^{-1} invariant. Thus, A is f $\pm$invariant iff $f(A) = A$.

Theorem 8.1. Let (X, f) be a reversible system with $\Phi_\pm : \beta \times X \to X$ the associated group system.

(a) A point is recurrent for the group system iff it is recurrent for (X, f) or for its reverse. That is,

(8.7) $\quad RECUR(\Phi_\pm) \;=\; RECUR(X,f) \cup RECUR(X,f^{-1})$.

The sets $RECUR(X,f), RECUR(X,f^{-1})$ and $RECUR(\Phi_\pm)$ are each $f \pm$ invariant as are their closures.

(b) The concepts of minimality, both for points of X and for subsets of X, agree for (X, f), for the reverse system (X, f^{-1}) and for the group system. Each minimal set is $f \pm$invariant as is the union $Min(X)$ and its closure.

(c) The prolongation relations on X satisfy the equations:

(8.8)
$$\mathcal{N}(f^{-1}) \;=\; (\mathcal{N}f)^{-1} \quad \text{and} \quad \mathcal{N}\Phi_\pm \;=\; \mathcal{N}f \cup \mathcal{N}(f^{-1})$$
$$\mathcal{N}^*(f^{-1}) \;=\; (\mathcal{N}^*f)^{-1} \quad \text{and} \quad \mathcal{N}^*\Phi_\pm \;=\; \mathcal{N}^*f \cup \mathcal{N}^*(f^{-1})$$

(c) The nonwandering sets of the three systems agree. That is,

(8.9) $\quad |\mathcal{N}^*f| \;=\; |\mathcal{N}^*(f^{-1})| \;=\; |\mathcal{N}^*\Phi_\pm|$.

(d) The proximality and regional proximality relations satisfy

(8.10)
$$PROX(\Phi_\pm) \;=\; PROX(f) \cup PROX(f^{-1}).$$
$$QPROX(\Phi_\pm) \;=\; QPROX(f) \cup QPROX(f^{-1}).$$

(e) (X, f) is transitive iff (X, f^{-1}) is transitive iff $\Phi_\pm$ is β^* transitive.

(f) The concepts of transitivity, point minimality, recurrent point minimality, distality, semidistality and equicontinuity agree for (X, f), for the reverse system (X, f^{-1}) and for the group system.

PROOF. (a): A point x is recurrent for $\Phi_\pm$ (or for (X, f)) iff $px = x$ for some $p \in \beta^*$ (resp. for some $p \in \beta^*_+$). So (8.3) implies (8.7). Furthermore, in either case, $pf(x) = f(px) = f(x)$ and $pf^{-1}(x) = f^{-1}(px) = f^{-1}(x)$. That is, p fixes the whole $\mathbb{Z}$ orbit. Thus,

(8.11) $\qquad Iso_x \;=\; Iso_{f^n(x)} \qquad$ for all $\quad n \in \mathbb{Z}$

It follows that each set of recurrent points is $f \pm$invariant which implies the same for the closure by Proposition 5.6(c).

(c),(d): For $A, B \subset X$,

(8.12)
$$N^{f^{-1}}(A, B) \;=\; N^f(B, A).$$
$$N^{\Phi_\pm}(A, B) \;=\; N^f(A, B) \cup (-N^{f^{-1}}(A, B))$$

It follows that $(x, y) \in \mathcal{N}(f^{-1})$ iff $(y, x) \in \mathcal{N}f$. The union is clearly contained in $\mathcal{N}\Phi_\pm$. If $N^{\Phi_\pm}(U, V)$ is nonempty for all neighborhoods of x and y then for a cofinal family either $N^f(U, V)$ or $N^{f^{-1}}(U, V)$ is nonempty. The same argument is used for $\mathcal{N}^*$. From the $\mathcal{N}^*$ equations it is obvious that the nonwandering sets agree.

(e): For proximality the result follows from the fact that $\beta = \beta_+ \cup \beta_-$. For regional proximality the argument is similar to the one given in (c) for prolongations.

In (b),(e) and (f), it suffices to compare (X, f) with the group system $\Phi_\pm$ since the group system for (X, f^{-1}) is, essentially, the same as $\Phi_\pm$ itself.

(b): If A is a closed f invariant set then so is $f(A) \subset A$. So if A is (X, f) minimal, it follows that $f(A) = A$ and so A is f ±invariant. A contains a closed subset B which is minimal for $\Phi_\pm$. Since B is β invariant, it follows that B is a closed f invariant subset of A and so $A = B$. Thus, A is $\Phi_\pm$ minimal. A similar argument shows that, conversely, a $\Phi_\pm$ minimal set is (X, f) minimal. A point is minimal iff it is contained in a minimal set. So the set result implies the point result.

(e): From (c) it is clear that f^{-1} is transitive if f is. By Theorem 7.4 they are in fact β_+^* transitive which certainly implies β^* transitivity for $\Phi_\pm$. Assume, conversely, that $\Phi_\pm$ is β^* transitive and let $U, V \subset X$ be opene. There exists $i \in N^{\Phi_\pm}(U, V)$. That is, $i \in \mathbb{Z}$ and $W = U \cap f^{-i}(V)$ is opene. $N^{\Phi_\pm}(W, W)$ is infinite by assumption and (8.12) implies it is symmetric about 0. Hence, both $\mathbb{Z}_+ \cap N(W, W)$ and $\mathbb{Z}_- \cap N(W, W)$ are infinite. But $j \in N(W, W)$ implies $i + j \in N(U, V)$. Thus, $\mathbb{Z}_+ \cap N(U, V)$ is infinite. This implies transitivity for (X, f).

(f): If (X, f) is point minimal then X decomposes into the disjoint union of f minimal subsets. As each of these is $\Phi_\pm$ minimal by part (b), it follows that $\Phi_\pm$ is a point minimal system. A similar application of (b) proves the converse.

If x is an $\Phi_\pm$ recurrent point with $K = \beta^*(x) = \beta x$ then x is a β^* transitive point for $\Phi_\pm | K$. By Theorem 6.3 $\Phi_\pm | K$ is β^* transitive. By part (e), $f | K$ is transitive. If f is recurrent point minimal then so is the restriction $f | K$. By Theorem 5.20(a), $f | K$ is minimal. By part (b) K is $\Phi_\pm$ minimal. Thus, (X, f) recurrent point minimal implies $\Phi_\pm$ is recurrent point minimal. Assume, conversely, that $\Phi_\pm$ is recurrent point minimal. By (8.7) any f recurrent point is recurrent for $\Phi_\pm$ and so is minimal by assumption. It is minimal for f by part (b) again. Since f is distal iff $f \times f$ is point minimal and f is semidistal iff $f \times f$ is recurrent point minimal (see Theorem 2.7(a),(c)), the last two results follow by applying what we have already proved to $f \times f$. If (X, f) is equicontinuous then because f is assumed surjective, we can apply Theorem 3.6 with $D = \{f^n : n \in \mathbb{Z}_+\}$. We obtain that the action of β_+ is equicontinuous and distal and that $\mathcal{E}_+(f) = \Phi^\#(\beta_+)$ is a compact subgroup of $\mathcal{H}(X)$. As the group contains all the iterates $\{f^n : n \in \mathbb{Z}\}$, it contains all of $\mathcal{E}(f) = \Phi^\#(\beta)$. It follows that $\Phi_\pm$ is equicontinuous. The converse is obvious. $\square$

REMARK 8.1. (a) If (X, f) is point transitive then it is transitive and so the reverse system (X, f^{-1}) is transitive. Unless we assume that X is metrizability we do not know whether (X, f^{-1}) is necessarily point transitive. We lack the leverage to apply to $TRANS$ the *Oxtoby Philosophy*: To find a point of a special type, it is often easiest to find a dense, G_δ set of such points.

(b) Using $f = M_1$, the translation action on $X = \beta$ defines a reversible system (X, f) which is not transitive (i.e. not transitive as an action of $(\beta_+, \mathbb{Z}_+, \beta_+^*)$). The group action $\Phi_\pm$ is point transitive and transitive, but is not β^* transitive.

Example (8) Noticeably missing from the list in Theorem 8.1 is almost distality. We now start from any infinite reversible, distal, minimal, metric system (Y, g) and construct a reversible metric system (X, f) which is an almost one-to-one lift of (Y, g), i.e. there exists an action map $\pi : (X, f) \to (Y, g)$ such that $Inj(\pi)$ is dense in X. By Proposition 4.10, such a map is irreducible and by Proposition 4.11 it is minimal and proximal. Hence, (X, f) is minimal. The special property of the map π built below is that it is asymptotic regarded as an action map from (X, f) to (Y, g) but not asymptotic regarded as an action map from (X, f^{-1}) to (Y, g^{-1}). Because it is the asymptotic lift of a distal system, (X, f) is almost distal by Theorem 2.9(d). So (X, f) is semidistal and its reverse is as well be Theorem 8.1(e). On the other hand, (X, f^{-1}) is a proximal, but not an asymptotic lift, so (X, f^{-1}) is not almost distal by Theorem 2.9(b) and (e).

If d is the metric on Y and $A \subset Y$ we define, as usual, $d(y, A) = inf\{d(y, y_1) : y_1 \in A\}$ so that $d(y, A) = 0$ iff y is in the closure of A.

For this construction begin with points $e, \tilde{e}$ with disjoint g orbits in the minimal set Y and choose increasing sequences of positive integers $\{n_i : i \in \mathbb{Z}_+\}$ and $\{m_i : i \in \mathbb{Z}_+\}$ such that $\{g^{-n_i}(e)\}$ and $\{g^{m_i}(\tilde{e})\}$ tend to e. Let $\{G_i\}$ be a pairwise disjoint sequence of open subsets of Y such that $g^{-n_i}(e) \in G_i$, e.g.

$$(8.13) \quad G_i = \{y \in Y : d(y, g^{-n_i}(e)) < d(y, \{g^{-n_j}(e)\} : j \neq i\}) \}.$$

For each $i \in \mathbb{Z}_+$ choose a positive integer J_i so that

$$(8.14) \quad j \geq J_i \implies m_j > n_i \text{ and } g^{-n_i + m_j}(\tilde{e}) \in G_i.$$

Define the dense G_δ subsets of Y

$$(8.15) \quad \begin{aligned} D_+ &= Y \setminus \{g^{-n}(e) : n = 0, 1, ...\} \quad \text{and} \\ D &= Y \setminus \{g^n(e) : n \in \mathbb{Z}\} = \\ &\bigcap_{n \in \mathbb{Z}} \{g^{-n}(Y \setminus \{e\})\} = \bigcap_{n \in \mathbb{Z}_+} \{g^n(D_+)\}. \end{aligned}$$

For $\epsilon = 0, 1$ the sets

$$(8.16) \quad A_\epsilon = \{g^{-n_i + m_k}(\tilde{e}) : k = 2j + \epsilon \text{ with } j \geq J_i \}$$

are disjoint closed subsets of D_+. With $I = [0,1]$, the unit interval in $\mathbb{R}$, define the continuous function $a : D_+ \to I$ by

$$(8.17) \qquad a(y) \;=\; d(y, A_0)/(d(y, A_0) + d(y, A_1)).$$

so that $a(y) = \epsilon$ for $y \in A_\epsilon$.

Let s be the shift homeomorphism on $I^{\mathbb{Z}}$ so that $s(z)_i = z_{i+1}$. Define the action map $Q : (D, g|D) \to (I^{\mathbb{Z}}, s)$ by

$$(8.18) \qquad Q(y)_i \;=\; a(g^i(y)) \qquad \text{for } i \in \mathbb{Z}.$$

Similarly, define the shift map s_+ on $I^{\mathbb{Z}_+ \setminus \{0\}}$ and the action map $Q_+ : (D_+, g|D_+) \to (I^{\mathbb{Z}_+\setminus\{0\}}, s_+)$.

We can regard the map Q as an invariant subset of $Y \times I^{\mathbb{Z}}$ and let X be the invariant subspace which is the closure of Q. Let f be the homeomorphism on X which is the restriction of $g \times s$ and let $\pi : (X, f) \to (Y, g)$ be the first coordinate projection map so that $\pi(y, z) = y$. Observe that continuity of Q and Q_+ imply that

$$(8.19) \quad \begin{aligned} (y,z) \in X \text{ and } y \in D &\implies z_i = g^i(y) \text{ for all } i \in \mathbb{Z}, \\ (y,z) \in X \text{ and } y \in D_+ &\implies z_i = g^i(y) \text{ for all } i \in \mathbb{Z}_+ \setminus \{0\}. \end{aligned}$$

This implies that $Inj(\pi) \supset Q$ and so π is an almost one-to-one map. For any $y \in Y$, the orbit $g^n(y)$ eventually enters and then remains in D_+. Hence, the action map $\pi : (X, f) \to (Y, g)$ is asymptotic. It follows that (X, f) is a reversible, minimal, almost distal system. Now for $\epsilon = 0, 1$ let $x^\epsilon = (e, z^\epsilon)$ be any limit point of the sequence $\{(g^{m_k}(\tilde{e}), Q(g^{m_k}(\tilde{e}))) : k = 2j + \epsilon, \ j \in \mathbb{Z}_+\}$. From the definition of the function a we see that

$$(8.20) \qquad z^\epsilon_{-n_i} \;=\; \epsilon \qquad \text{for all } i \in \mathbb{Z}_+.$$

Thus, (x^0, x^1) is a pair in R_π which is not asymptotic in the negative direction. Hence, the action map $\pi : (X, f^{-1}) \to (Y, g^{-1})$ is not asymptotic. Since it is almost one-to-one it is still proximal and so (X, f) is not almost distal.

Example (9) Beginning with a reversible system (Y, g) on an infinite, zero-dimensional metrizable space Y with Y minimal, we will construct reversible systems (X, f) and (Z, h) and asymptotic, continuous action maps $\pi : X \to Z$ and $\rho : Z \to Y$ such that the composition $\rho \circ \pi : X \to Y$ is almost one-to-one but not asymptotic. Both Z and X are minimal and if (Y, g) is distal then (Z, h) is almost distal by Theorem 2.9(d). However, by Theorem 2.9(b) and (e), (X, f) is not almost distal. As asymptotic maps are almost distal, the composition $\rho \circ \pi$ is semidistal by Proposition 1.7(d) and so (X, f) is semidistal.

For (Y, g) and $y \in Y$ we describe the *Orbit Doubling Construction*.

Begin with a strictly decreasing sequence $\{U^n : n \in \mathbb{Z}_+\}$ of clopen neighborhoods of y with $U^0 = Y$ and partition each $U^n \setminus U^{n+1}$ into two nonempty clopen pieces A_0^n and A_1^n. For $\epsilon = 0, 1$ let

(8.21) $$A_\epsilon \quad = \quad \{y\} \cup \bigcup_n A_\epsilon^n.$$

A_ϵ is closed in Y with interior $A_\epsilon \setminus \{y\}$. Furthermore,

(8.22) $$A_0 \cap A_1 \quad = \quad \{y\} \quad \text{and} \quad A_0 \cup A_1 \quad = \quad Y.$$

With $J = \{0, 1\}$ let s be the shift homeomorphism on $J^{\mathbb{Z}}$ so that $s(a)_n = a_{n+1}$. Denote by (Z, h) the subsystem for the product system $(Y \times J^{\mathbb{Z}}, g \times s)$ with:

(8.23) $$Z \quad =_{def} \quad \{(w, a) \in Y \times J^{\mathbb{Z}} : g^n(w) \in A_{a_n} \text{ for all } n \in \mathbb{Z}\}.$$

Let $\rho : Z \to Y$ be the the restriction of the projection map. Observe that the fibers of ρ are all singletons except above the points in the orbit of y. $\rho^{-1}\{g^i(y)\}$ consists of two points $(g^i(y), a)$ and $(g^i(y), b)$ with $a_n = b_n$ for all $n \neq -i$. In particular, the pair (a, b) is asymptotic for $(J^{\mathbb{Z}}, s)$ and for its reverse system. Thus, ρ is an asymptotic map. Let $z^0 = (y, a)$ and $z^1 = (y, b)$ with $a_0 = 0, b_0 = 1$. Now we define (X, f) a subsystem of $(Z \times J^{\mathbb{Z}}, h \times s)$ by a similar orbit doubling of z^0 with an extra contrivance. Let $\{n_k\}$ be an increasing sequence of positive integers such that

(8.24) $$Lim_{n_k} \quad h^{n_k}(z^0) \quad = \quad Lim_{n_k} \quad h^{n_k}(z^1) \quad = \quad z^0.$$

In defining the analogues B_ϵ of the A_ϵ above we can make sure that

(8.25) $$h^{n_k}(z^\epsilon) \quad \in \quad B_\epsilon \quad \text{for all} \quad k \in \mathbb{Z}_+.$$

because the two convergent subsequences are disjoint.

Let $x^1 = (z^1, c)$ denote the unique point of X with $\pi(x^1) = z^1$, where $\pi : X \to Z$ is the restriction of the projection map. Let $x^{00} = (z^0, a)$ and $x^{01} = (z^0, b)$ with $a_0 = 0, b_0 = 1$. By (8.25) and the analogue of (8.23) we see that for all $k \in \mathbb{Z}_+$

(8.26) $$a_{n_k} \quad = \quad b_{n_k} \quad = \quad 0 \quad \text{and} \quad c_{n_k} \quad = \quad 1.$$

It follows that the pairs (x^{00}, x^1) and (x^{01}, x^1) are not asymptotic for (X, f). However, x^{00}, x^{01} and x^1 are all mapped to y by $\rho \circ \pi$.

It is interesting to note that there exist semidistal actions which cannot be obtained by lifting from the trivial action via a tower of distal and asymptotic maps. For example, in Auslander (1983) Chapter 1 a "non-homogeneous" minimal set is constructed by modifying an example due to E. E. Floyd. It is obtained as an almost one-to-one lifting $\pi : (X, f) \to (Y, g)$ with (Y, g) distal (in fact, equicontinuous). It can be shown that for this example

(8.27) $$ASYMP(f) \cap R_\pi \quad = \quad 1_X \quad = \quad RECUR^2(f) \cap R_\pi.$$

Hence, π is semidistal but not almost distal. Suppose (X, f) is decomposed in such manner. By taking the pointed joining of each system in the tower with the system Y we can assume that $\pi : (X, f) \to (Y, g)$ is the composition of the tower. Now suppose $\pi = \delta \circ \epsilon$ with neither map a homeomorphism. By Proposition 4.10(d) both δ and ϵ are irreducible and hence are proximal. So neither is distal. On the other hand, by (8.27) no off-diagonal element of $R_\epsilon \subset R_\pi$ is asymptotic. Hence, ϵ is not an asymptotic map.

We conclude with an entropy result in the spirit of the work Blanchard et al. (2002), we refer the reader to this paper for more details. For a detailed treatment of the the subject of entropy pairs see Glasner (2003).

Theorem 8.2. A minimal, semidistal system (X, f) has topological entropy zero.

PROOF. Let μ be an invariant probability measure for the system. As shown in Glasner (1997) the set Λ_μ, defined to be the closure of the set of entropy pairs for μ, is a closed, invariant, transitive subset of the product $X \times X$ which contains the diagonal 1_X. Since (X, f) is semidistal, the product $(X \times X, f \times f)$ and so the subsystem Λ_μ is semidistal as well by Proposition 1.9 and Corollary 1.8. By Theorem 5.21(a), Λ_μ is minimal and so equals the diagonal. This implies that the entropy $h_\mu(X, f) = 0$. As this is true for every invariant measure μ, the variational principle, see e.g. Misiurewicz (1976), implies that the topological entropy is zero. □

Table

We conclude with the following table which summarizes some of the main results (at least) for minimal $\mathbb{Z}$ systems (with $S = \beta^*(\mathbb{Z}_+)$ or $S = \beta^*(\mathbb{Z}_+ \cup \mathbb{Z}_-)$). PE stands for the property "proximal is an equivalence relation" which is equivalent to $PROX = LPROX$.

	Distal	Almost distal	Semi distal	PE	PI
Definition	$PROX(f) = \mathbf{1}_X$	$PROX(f) = ASYMP(f)$	$PROX(f) \cap RECUR^2(f) = \mathbf{1}_X$	$PROX(f) = LPROX(f)$	Proximal factor of a tower of proximal and isometric extensions
Enveloping semigroup characterization	$\mathcal{E}(f)$ is a group	$\mathcal{A}(f)$ is minimal	Each idempotent in $\mathcal{A}(f)$ is minimal	$\mathcal{E}(f)$ contains a unique minimal ideal	
$X \times X$ characterization	Each pair is minimal	Orbit closure = orbit ∪ a minimal set	Each recurrent pair is minimal	Each orbit closure contains a unique minimal set	Transitive subsets with a dense set of minimal points are minimal
S characterization	S distal	S almost distal	S semi distal	$[Min(S)]$ almost distal	$[Min(S)]$ semi distal
Closure under factors	✓	✓	✓	✓	✓
Closure under extensions of the same kind	✓	✠	✓	✠	✓
Zero entropy	✓	✓	✓	✠	✠
f^{-1}	✓	✠	✓	✠	✓
Typical examples	$(x, y) \mapsto (x + \alpha, y + x)$ on $\mathbb{T}^2$	Sturmian systems, Morse system (one-sided)	Shapiro's examples	Toeplitz, Proximal extensions of distal systems	Substitutions of constant length

TABLE 1. Distality properties for minimal systems

Bibliography

[1] E. Akin (1993) **The general topology of dynamical systems** Amer. Math. Soc., Providence.

[2] E. Akin (1997) **Recurrence in topological dynamics: Furstenberg families and Ellis actions** Plenum Publishing Co., New York.

[3] E. Akin (2004) *Lectures on Cantor and Mycielski sets for dynamical systems* Contemporary Math. AMS **356**:21-79.

[4] E. Akin and E. Glasner (1998) *Topological ergodic decomposition and homogeneous flows* Contemporary Math. AMS **215**:43-52

[5] E. Akin and E. Glasner (2001) *Residual properties and almost equicontinuity* J. d'Anal. Math.**84**:243-286.

[6] J. Auslander (1964) *Generalized recurrence in dynamical systems*, **Contributions to Differential Equations**, vol. 3, John Wiley, New York :55-74.

[7] J. Auslander (1988) **Minimal flows and their extensions** North Holland, Amsterdam.

[8] J. Auslander (2004) *Proximality and regional proximality in minimal flows* Illinois J. Math. **45**:665-673.

[9] J. Auslander and H. Furstenberg (1994) *Product recurrence and distal points* Trans AMS.**343**:221-232.

[10] J. Auslander and E. Glasner (1977) *Distal and highly proximal extensions of minimal flows* Indiana Univ. Math. J. **26**:731-749.

[11] J. Auslander and E. Glasner (2002) *The distal order of a minimal flow* Israel J. Math. **127** :61-80.

[12] J. Auslander and J.A. Yorke (1980) *Interval maps, factors of maps and chaos* Tohoku Math. J. **32**:177-188.

[13] L. Auslander, L. Green, and F. Hahn (1963). **Flows on homogeneous spaces**. With the assistance of L. Markus and W. Massey, and an appendix by L. Greenberg. Annals of Mathematics Studies, **No. 53** Princeton University Press, Princeton, N.J.

[14] F. Blanchard, E. Glasner, S. Kolyada, and A. Maass (2002) *On Li-Yorke pairs* J. Reine Angew. Math. **No. 547**:51-68

[15] F. Blanchard, B. Host, and A. Maass (2000) *Topological complexity* Ergodic Th. & Dynam. Sys. **20**:641-662.

[16] I. U. Bronstein (1977) *A characteristic property of PD-extensions* Bull. Akad. Stiince RSS Moldoven **3**:11-15.

[17] R. Burckel (1970) **Weakly almost periodic functions on semigroups** Gordon and Breach, New York.

[18] J. P. Clay (1963), *Proximity relations in transformation groups* Trans. Amer. Math. Soc. **108**:88-96.

[19] R. Devaney (1989) **Chaotic dynamical systems** 2nd ed. Addison-Wesley, New York.

[20] D. Ellis, R. Ellis and M. Nerurkar (2000) *The topological dynamics of semigroup actions* Trans. AMS **353**:1279-1320.

[21] R. Ellis (1957_1) *A note on the continuity of the inverse* Proc. AMS **8**:37-373.

[22] R. Ellis (1957_2) *Locally compact transformation groups* Duke Math. Jour. **24**:119-126.

[23] R. Ellis (1969) **Lectures on topological dynamics** W. A. Benjamin, New York.

[24] R. Ellis, E. Glasner and L. Shapiro (1975) *Proximal-isometric (PI) flows.* Advances in Math. **17**:213–260

[25] H. Furstenberg (1967) *Disjointness in ergodic theory, minimal sets and a problem in diophantine approximation* Math. Systems Th. **1**:1-55.

[26] H. Furstenberg (1981) **Recurrence in ergodic theory and combinatorial number theory** Princeton Univ. Press, Princeton.

[27] E. Glasner (1976) **Proximal flows** Lect. Notes in Math #517, Springer-Verlag, Berlin.

[28] E. Glasner (1980) *Divisible properties and the Stone-Čech compactification* Canad. J. Math. **32**:993-1007

[29] E. Glasner (1990) *A topological version of a theorem of Veech and almost simple flows* Ergod. Theory Dynam. Systems **10**:463-482

[30] E. Glasner (1997) *A simple characterization of the set of μ-entropy pairs and applications* Israel J. Math. **102**:13-27.

[31] E. Glasner (2003) **Ergodic theory via joinings** Amer. Math. Soc., Providence.

[32] E. Glasner (2005) *Topological weak mixing and quasi-Bohr systems* Israel J. Math. **128**:277-304.

[33] E. Glasner and D. Maon (1989) *Rigidity in topological dynamics* Ergod. Th. & Dynam. Sys. **9**:309-320.

[34] N. Hindman and D. Strauss (1998) **Algebra in the Stone-Čech compactification** Walter de Gruyter, Berlin.

[35] W. Huang and X. Ye (2002) *Devaney's chaos or 2-scattering implies Li-Yorke's chaos* Topology and its Applications **117**:259-272.

[36] J. Kelley (1955) **General Topology** Van Nostrand, Princeton.

[37] H. B. Keynes (1967) *A study of the proximal relation in coset transformation groups* Trans. Amer. Math. Soc. **128**:389-402.

[38] S. Kolyada, L. Snoha and S. Trofimchuk (2001) *Noninvertible minimal maps* Fundamenta Mathematicae **168**: 141-163.

[39] K. Kuratowski (1966) **Topology** Academic Press, New York.

[40] T.Y. Li and J.A. Yorke (1975) *Period three implies chaos* Amer. Math. Monthly **82**:985-992.

[41] N. Markley (1972) *F- minimal sets* Trans. Amer. Math. Soc. **163**:85-100.

[42] D. C. McMahon (1976) *Weak mixing and a note on the structure theorem for minimal transformation groups* Illinois J. Math. **20**:186-197.

[43] M. Misiurewicz (1976) *A short proof of the variational principle for a $\mathbf{Z}_+^N$ action on a compact space* Astérisque **40**:147-187.

[44] J. Oxtoby (1980) **Measure and category**(2nd Ed.), Springer-Verlag, Berlin.

[45] R. Peleg (1972) *Weak disjointness of transformation groups* Proc. Amer. Math. Soc. **33**:165-170.

[46] W. Ruppert (1984) **Compact semitopological semigroups: an intrinsic theory** Lect. Notes in Math #1079, Springer-Verlag, Berlin.

[47] L. Shapiro (1970) *Proximality in minimal transformation groups* Proc. Amer. Math. Soc. **26**:521-525.

BIBLIOGRAPHY

[48] P. Tondeur (1965) **Introduction to Lie groups and transformation groups** Lect. Notes in Math #7, Springer-Verlag, Berlin.

[49] J. de Vries (1993) **Elements of topological dynamics** Kluwer Academic Publishers, Dordrecht.

[50] W. A. Veech (1970) *Point-distal flows* Amer. J. Math. **92**:205-242.

[51] W. A. Veech, (1977) *Topological dynamics* Bull. Amer. Math. Soc. **83**:775-830

[52] B. Weiss (2000) *A survey of generic dynamics* in **Descriptive set theory and dynamical systems** (ed. M. Foreman, A.S. Kechris, A. Louveau and B. Weiss) London Math. Soc. Lecture Notes # 277, Cambridge U Press, Cambridge :273-291.

[53] J.C.S.P. van der Woude (1985) *Characterizations of (H)PI-extensions* Pacific J. Math. **120**:453-467.

[54] J.C.S.P. van der Woude (1986) **Topological dynamics**, CWI Tracts **22**, Center for Mathematics and Computer Science, Amsterdam.

Index

$>_R, >_L$, 21
2^X, 110
2^π, 114

$[A]$, 20, 41
$[[\mathcal{A}]]$, 68
$\mathcal{A}_+(f)$, 135
action, 9
    almost distal action, 14
    asymptotic action, 13
    bizarre action, 85
    classical action, 60
    densely continuous action, 41
    distal action, 14
    Ellis action, 19
    equicontinuous action, 37
    faithful action, 11
    identity action, 38, 56
    product action, 11
    proximal action, 13
    semidistal action, 14
    topological action, 35
    translation action, 9
    trivial action, 9, 13
action map, 2, 10
    almost distal action map, 14
    asymptotic action map, 14
    distal action map, 14
    minimal action map, 31
    proximal action map, 14
    semidistal action map, 14
    trivial action map, 10
Ad_f, 37
adherence semigroup, 135
adjoint action, 97
adjoint isotropy set, 98
adjoint prolongation relation, 95
adjoint recurrent point, 98
adjoint transitive, 95
adjoint transitive point, 98

S^* adjoint transitive point, 98
almost distal, 3
    almost distal action, 14
    almost distal action map, 14
almost one-to-one map, 56
$ASYMP$, 3, 13, 18
asymptotic, 2, 13
    asymptotic action, 13
    asymptotic action map, 14

$\mathcal{B}$, 68
$\mathcal{B}^u(G)$, 62
β, β^*, 135
$\beta_u G$, 62
$\beta_u^* G$, 63
$\beta_u \mathbb{R}_+, \beta_u^* \mathbb{R}_+$, 63
$\beta_u \mathbb{Z}_+, \beta_u^* \mathbb{Z}_+$, 63
$Bdry$, 71
bizarre action, 85
bounded subset, 59

$\mathcal{C}(X,X), \mathcal{C}_s(X,X)$, 36
cancelation subsemigroup, 10
capturing property, 49, 53, 91
cascade, 135
$Cent(\Phi, x)$, 95
center, 32, 48
    min-center, 32, 48
Φ centralizer, 95
classical action, 60
    surjective classical action, 66
classical Ellis semigroup, 59
a classical Ellis semigroup satisfies the
    Surjection Condition, 66
co-ideal, 10
$Comp$, 37
compact-open topology, 36
composition results, 17, 30, 54, 56
Convention (for invariance and
    recurrence), 65

Δ_X^n, 107
densely continuous, 41, 60
Dichotomy Theorem, 116
disjoint systems, 81
  weakly disjoint systems, 81
distal, 2, 3
  almost distal, 3
  distal action, 14
  distal action map, 14
  distal point, 32, 120
  semidistal, 3
divisible property, 68
dynamical system, 1
  adjoint transitive system, 95
  adjoint point transitive system, 98
  adjoint weak mixing system, 102
  $\mathcal{F}$ transitive system, 76
  group system for a homeomorphism, 136
  Li-Yorke chaotic system, 3
  map dynamical system, 135
  metric system, 1
  minimal system, 1, 11
  mixing system, 76, 102
  orbit minimal system, 25
  PI system, 120
  point distal system, 2
  point minimal system, 25
  point transitive system, 1, 11, 25
  point weak mixing system, 56
  recurrent point minimal system, 25
  reverse system, 1, 136
  reversible system, 1, 136
  S^* adjoint transitive system, 95
  S^* adjoint point transitive system, 98
  S^* transitive system, 76
  scattering system, 81
  strongly Li-Yorke chaotic system, 3
  subsystem, 2, 10
  totally recurrent system, 25
  transitive system, 1, 76
  weak mixing system, 2, 81, 130
dual family, 68

$\mathcal{E}(f), \mathcal{E}_+(f)$, 135
Ellis action, 19
Ellis-Numakura Lemma, 208
Ellis semigroup, 19
  classical Ellis semigroup, 59
enveloping semigroup, 11, 135
entropy, 142
equicontinuous, 36
equicontinuous action, 37

Ev, 37
Examples, 62, 63, 86, 103, 128, 139, 140

$\Phi^\#, \Phi_\#, \Phi^p, \Phi_x$, 9
$\Phi_\pm$, 136
faithful action, 11
family, 67
  dual family, 68
  family generated by $\mathcal{A}$, 68
  proper family, 67
filter, 68
filterdual, 68
$Fin(A)$, 95
fixed point, 11
$Foc_{(x_1,x_2)}$, 11
focus set, 11
Fubini's Theorem, 109
Furstenberg Intersection Lemma, 101, 130

$G^{-1}A$, 77, 108
Gelfand construction, 61
group system for a homeomorphism, 136

$\mathcal{H}(X)$, 36
$H(\mathcal{F})$, 69
hereditary upwards, 67
highly proximal map, 57
hitting time set, 69
homogeneous space, 92
homomorphism, 10
  monoid homomorphism, 10
hull of $\mathcal{F}$, 69

$Id(A)$, 20
ideal, 10
  closed ideal generated by C (= $[C]$), 41
  principal ideal, 10
  two-sided ideal, 12
idempotent, 10, 20
  maximal idempotent, 22
  minimal idempotent, 21
identity action, 38, 56
$Inj(\pi)$, 56
Intersection Condition, 121
Inv, 37
invariance Convention, 65
invariant subset, 2, 10, 135
  closed invariant subset generated by A (= $[A]$), 20, 41
  minus invariant subset, 63
inverse limit, 60

surjective inverse limit, 60
irreducible map, 56
Iso_x, 11
Iso_x^{adj}, 98
Iso_X, 124
isotropy set, 11
    adjoint isotropy set, 98

join of two families, 68

$k\mathcal{B}$, 68
$k\mathcal{F}$, 68
$k\mathcal{P}_+$, 68
Kuratowski-Mycielski Theorem, 118
Kuratowski-Ulam Theorem, 109

λ_π, 14
λ_π^n, 107
$\lambda_{\pi,\rho}$, 105
$LI - YORKE$, 3
Li-Yorke chaotic, 3
    strongly Li-Yorke chaotic, 3, 120
$LPROX$, 49

M^p, 9
map dynamical system, 135
maximal idempotent, 22
measure, 104, 142
    measure of full support, 104
$Min(X), Min(S)$, 12, 20
$[Min(X)], [Min(S)]$, 32
min-center, 32, 48
minimal action map, 31
minimal ideal, 12
minimal idempotent, 21
minimal point, 11
minimal subset, 2, 11, 19
minimal system, 11
minus invariant subset, 63
mixing, 76, 102
monoid, 8
    monoid action, 9
    monoid extension, 9
    monoid homomorphism, 10
Mycielski subset, 118

$\mathcal{N}^n$, 108
$\mathcal{N}(V)$, 37
$\mathcal{N}\Phi$, 70
$\mathcal{N}_\mathcal{F}\Phi$, 70
$\mathcal{N}_{adj}\Phi, \mathcal{N}_{\mathcal{F}adj}\Phi$, 95
$\mathcal{N}^*\Phi$, 72
$\mathcal{N}_{adj}^*\Phi$, 95
$|\mathcal{N}^*\Phi|$, 73

$N^\Phi(A,B)$, 69
$N_F^\Phi(U,V)$, 94
$N^f(U,V)$, 136
Numakura, 20
nonwandering point, 73
nonwandering set, 73

$\omega f(x)$, 1, 64
one-point compactification, 64
opene set, 69
orbit, 10
orbit closure, 1, 60
Orbit Doubling Construction, 140
orbit minimal, 25
orbit sequence, 1
Oxtoby Philosophy, 139

$\mathcal{P}_+$, 68
PI system, 120
point distal, 2
point minimal, 25
point transitive, 1, 11, 25, 75
    point transitive subset, 2
    S^* point transitive, 75
point weak mixing, 56
product action, 11
product recurrent point, 33
projection semigroup, 39
prolongation, 70
    adjoint prolongation, 95
proper family, 67
$PROX$, 3, 13, 23
    $PROX_H$, 45
    $LPROX$, 49
    $QPROX$, 74
    $QPROX^n$, 108
$PROX \wedge RECUR$, 13, 23
$PROX^n$, 107
proximal, 2, 13
    highly proximal action map, 57
    proximal action, 13
    proximal action map, 14
    proximal cell, 2
pullback, 10, 105

$QPROX$, 74
$QPROX^n(\Phi), QPROX^n(\pi)$, 108

R_π, 2, 14
R_π^n, 107
$R_{\pi,\rho}$, 105
$RECUR$, 13, 23
    $RECUR_H$, 45
$[RECUR]$, 32, 48

$RECUR^2$, 3
$RECUR^n$, 13
recurrence Convention, 65
recurrent point, 2, 11
  adjoint recurrent point, 98
  product recurrent point, 33
recurrent point minimal, 25
regional proximality relation, 74
relation, 69
  closed relation, 70
  inverse relation, 69
residual property, 82
restriction, 10
reverse system, 1, 136
reversible system, 1, 136
right-invariant uniformity, 62

S^+, 9
scattering, 81
scrambled set, 3
  strongly scrambled set, 3, 120
semidistal, 3
  semidistal action, 14
  semidistal action map, 14
semigroup, 9
  adherence semigroup, 135
  classical Ellis semigroup, 59
  Ellis semigroup, 19
  enveloping semigroup 11, 135
  projection semigroup, 39
  semigroup action, 9
  semigroup homomorphism, 10
  subsemigroup, 10
  topological semigroup, 35
separable, 59, 60
shadow diagram, 113
$sLI - YORKE$, 3
Stone-Čech compactification, 4, 62
stronger property, 82
strongly Li-Yorke chaotic, 3, 120
strongly scrambled set, 3, 120
submonoid, 10
subsemigroup, 10
  cancelation subsemigroup, 10
subsystem, 2, 17
Surjection Condition, 66
surjective classical action, 66
surjective inverse limit, 60
surjective inverse system, 60
syndetic proximality relation, 49
syndetic subgroup, 92

$\mathcal{T}^\Phi, \overline{\mathcal{T}}^\Phi$, 100, 131

$\mathcal{T}^\Phi_{adj}, \overline{\mathcal{T}}^\Phi_{adj}$, 100
topological action, 35
topological entropy, 142
topological semigroup, 35
totally recurrent, 25
trace, 68
$TRANS$, 11, 25, 136
  $TRANS_f$, 1
  $TRANS^{adj}, TRANS^{adj}_{S^*}$, 98
  $TRANS_H$, 45
  $TRANS_{S^*}$, 75
transitive system, 76
  adjoint point transitive system, 98
  adjoint transitive system, 95
  $\mathcal{F}$ transitive system, 76
  S^* adjoint point transitive system, 98
  S^* adjoint transitive system, 95
  S^* transitive system, 76
translation action, 9
trivial action, 9
trivial action map, 10

$\mathcal{U}_G$, 62
unbounded subset, 59

Veech shadow diagram, 113

weakly disjoint, 81
weak mixing, 2, 81, 130
  adjoint weak mixing, 102
  weak mixing of all orders, 2, 81
  point weak mixing, 56

Editorial Information

To be published in the *Memoirs*, a paper must be correct, new, nontrivial, and significant. Further, it must be well written and of interest to a substantial number of mathematicians. Piecemeal results, such as an inconclusive step toward an unproved major theorem or a minor variation on a known result, are in general not acceptable for publication.

Papers appearing in *Memoirs* are generally at least 80 and not more than 200 published pages in length. Papers less than 80 or more than 200 published pages require the approval of the Managing Editor of the Transactions/Memoirs Editorial Board.

As of May 31, 2008, the backlog for this journal was approximately 17 volumes. This estimate is the result of dividing the number of manuscripts for this journal in the Providence office that have not yet gone to the printer on the above date by the average number of monographs per volume over the previous twelve months, reduced by the number of volumes published in four months (the time necessary for preparing a volume for the printer). (There are 6 volumes per year, each usually containing at least 4 numbers.)

A Consent to Publish and Copyright Agreement is required before a paper will be published in the *Memoirs*. After a paper is accepted for publication, the Providence office will send a Consent to Publish and Copyright Agreement to all authors of the paper. By submitting a paper to the *Memoirs*, authors certify that the results have not been submitted to nor are they under consideration for publication by another journal, conference proceedings, or similar publication.

Information for Authors

Memoirs are printed from camera copy fully prepared by the author. This means that the finished book will look exactly like the copy submitted.

Initial submission. The AMS uses Centralized Manuscript Processing for initial submissions. Authors should submit a PDF file using the Initial Manuscript Submission form found at www.ams.org/cgi-bin/peertrack/submission.pl, or send one copy of the manuscript to the following address: Centralized Manuscript Processing, MEMOIRS OF THE AMS, 201 Charles Street, Providence, RI 02904-2294 USA. If a paper copy is being forwarded to the AMS, indicate that it is for it Memoirs and include the name of the corresponding author, contact information such as email address or mailing address, and the name of an appropriate Editor to review the paper (see the list of Editors below).

The paper must contain a *descriptive title* and an *abstract* that summarizes the article in language suitable for workers in the general field (algebra, analysis, etc.). The *descriptive title* should be short, but informative; useless or vague phrases such as "some remarks about" or "concerning" should be avoided. The *abstract* should be at least one complete sentence, and at most 300 words. Included with the footnotes to the paper should be the 2000 *Mathematics Subject Classification* representing the primary and secondary subjects of the article. The classifications are accessible from www.ams.org/msc/. The list of classifications is also available in print starting with the 1999 annual index of *Mathematical Reviews*. The Mathematics Subject Classification footnote may be followed by a list of *key words and phrases* describing the subject matter of the article and taken from it. Journal abbreviations used in bibliographies are listed in the latest *Mathematical Reviews* annual index. The series abbreviations are also accessible from www.ams.org/publications/. To help in preparing and verifying references, the AMS offers MR Lookup, a Reference Tool for Linking, at www.ams.org/mrlookup/.

Electronically prepared manuscripts. The AMS encourages electronically prepared manuscripts, with a strong preference for $\mathcal{A}_{\mathcal{M}}\mathcal{S}$-LaTeX. To this end, the Society has prepared $\mathcal{A}_{\mathcal{M}}\mathcal{S}$-LaTeX author packages for each AMS publication. Author packages include instructions for preparing electronic manuscripts, samples, and a style file that generates the particular design specifications of that publication series. Though $\mathcal{AMS}$-LaTeX is the highly preferred format of TeX, author packages are also available in $\mathcal{AMS}$-TeX.

Authors may retrieve an author package from the AMS website starting from www.ams.org/tex/ or via FTP to ftp.ams.org (login as anonymous, enter username as password, and type cd pub/author-info). The *AMS Author Handbook* and the *Instruction Manual* are available in PDF format following the author packages link from www.ams.org/tex/. The author package can also be obtained free of charge by sending email to tech-support@ams.org (Internet) or from the Publication Division, American Mathematical Society, 201 Charles St., Providence, RI 02904-2294, USA. When requesting an author package, please specify $\mathcal{AMS}$-LaTeX or $\mathcal{AMS}$-TeX and the publication in which your paper will appear. Please be sure to include your complete mailing address.

After acceptance. The final version of the electronic file should be sent to the Providence office (this includes any TeX source file, any graphics files, and the DVI or PostScript file) immediately after the paper has been accepted for publication.

Before sending the source file, be sure you have proofread your paper carefully. The files you send must be the EXACT files used to generate the proof copy that was accepted for publication. For all publications, authors are required to send a printed copy of their paper, which exactly matches the copy approved for publication, along with any graphics that will appear in the paper.

Accepted electronically prepared files can be submitted via the web at www.ams.org/submit-book-journal/, sent via FTP, or sent on CD-Rom or diskette to the Electronic Prepress Department, American Mathematical Society, 201 Charles Street, Providence, RI 02904-2294 USA. TeX source files, DVI files, and PostScript files can be transferred over the Internet by FTP to the Internet node ftp.ams.org (130.44.1.100). When sending a manuscript electronically via CD-Rom or diskette, please be sure to include a message identifying the paper as a Memoir.

Electronically prepared manuscripts can also be sent via email to pub-submit@ams.org (Internet). In order to send files via email, they must be encoded properly. (DVI files are binary and PostScript files tend to be very large.)

Electronic graphics. Comprehensive instructions on preparing graphics are available at www.ams.org/jourhtml/. A few of the major requirements are given here.

Submit files for graphics as EPS (Encapsulated PostScript) files. This includes graphics originated via a graphics application as well as scanned photographs or other computer-generated images. If this is not possible, TIFF files are acceptable as long as they can be opened in Adobe Photoshop or Illustrator. No matter what method was used to produce the graphic, it is necessary to provide a paper copy to the AMS.

Authors using graphics packages for the creation of electronic art should also avoid the use of any lines thinner than 0.5 points in width. Many graphics packages allow the user to specify a "hairline" for a very thin line. Hairlines often look acceptable when proofed on a typical laser printer. However, when produced on a high-resolution laser imagesetter, hairlines become nearly invisible and will be lost entirely in the final printing process.

Screens should be set to values between 15% and 85%. Screens which fall outside of this range are too light or too dark to print correctly. Variations of screens within a graphic should be no less than 10%.

Inquiries. Any inquiries concerning a paper that has been accepted for publication should be sent to memo-query@ams.org or directly to the Electronic Prepress Department, American Mathematical Society, 201 Charles St., Providence, RI 02904-2294 USA.

Editors

This journal is designed particularly for long research papers, normally at least 80 pages in length, and groups of cognate papers in pure and applied mathematics. Papers intended for publication in the *Memoirs* should be addressed to one of the following editors. The AMS uses Centralized Manuscript Processing for initial submissions to AMS journals. Authors should follow instructions listed on the Initial Submission page found at www.ams.org/memo/memosubmit.html.

Algebra to ALEXANDER KLESHCHEV, Department of Mathematics, University of Oregon, Eugene, OR 97403-1222; email: ams@noether.uoregon.edu

Algebraic geometry and its application to MINA TEICHER, Emmy Noether Research Institute for Mathematics, Bar-Ilan University, Ramat-Gan 52900, Israel; email: teicher@macs.biu.ac.il

Algebraic geometry to DAN ABRAMOVICH, Department of Mathematics, Brown University, Box 1917, Providence, RI 02912; email: amsedit@math.brown.edu

Algebraic topology to ALEJANDRO ADEM, Department of Mathematics, University of British Columbia, Room 121, 1984 Mathematics Road, Vancouver, British Columbia, Canada V6T 1Z2; email: adem@math.ubc.ca

Combinatorics to JOHN R. STEMBRIDGE, Department of Mathematics, University of Michigan, Ann Arbor, Michigan 48109-1109; email: FRS@umich.edu

Complex analysis and harmonic analysis to ALEXANDER NAGEL, Department of Mathematics, University of Wisconsin, 480 Lincoln Drive, Madison, WI 53706-1313; email: nagel@math.wisc.edu

Differential geometry and global analysis to LISA C. JEFFREY, Department of Mathematics, University of Toronto, 100 St. George St., Toronto, ON Canada M5S 3G3; email: jeffrey@math.toronto.edu

Dynamical systems and ergodic theory and complex anaysis to YUNPING JIANG, Department of Mathematics, CUNY Queens College and Graduate Center, 65-30 Kissena Blvd., Flushing, NY 11367; email: Yunping.Jiang@qc.cuny.edu

Functional analysis and operator algebras to DIMITRI SHLYAKHTENKO, Department of Mathematics, University of California, Los Angeles, CA 90095; email: shlyakht@math.ucla.edu

Geometric analysis to WILLIAM P. MINICOZZI II, Department of Mathematics, Johns Hopkins University, 3400 N. Charles St., Baltimore, MD 21218; email: trans@math.jhu.edu

Geometric analysis to MARK FEIGHN, Math Department, Rutgers University, Newark, NJ 07102; email: feighn@andromeda.rutgers.edu

Harmonic analysis, representation theory, and Lie theory to ROBERT J. STANTON, Department of Mathematics, The Ohio State University, 231 West 18th Avenue, Columbus, OH 43210-1174; email: stanton@math.ohio-state.edu

Logic to STEFFEN LEMPP, Department of Mathematics, University of Wisconsin, 480 Lincoln Drive, Madison, Wisconsin 53706-1388; email: lempp@math.wisc.edu

Number theory to JONATHAN ROGAWSKI, Department of Mathematics, University of California, Los Angeles, CA 90095; email: jonr@math.ucla.edu

Partial differential equations to GUSTAVO PONCE, Department of Mathematics, South Hall, Room 6607, University of California, Santa Barbara, CA 93106; email: ponce@math.ucsb.edu

Partial differential equations and dynamical systems to PETER POLACIK, School of Mathematics, University of Minnesota, Minneapolis, MN 55455; email: polacik@math.umn.edu

Probability and statistics to RICHARD BASS, Department of Mathematics, University of Connecticut, Storrs, CT 06269-3009; email: bass@math.uconn.edu

Real analysis and partial differential equations to DANIEL TATARU, Department of Mathematics, University of California, Berkeley, Berkeley, CA 94720; email: tataru@math.berkeley.edu

All other communications to the editors should be addressed to the Managing Editor, ROBERT GURALNICK, Department of Mathematics, University of Southern California, Los Angeles, CA 90089-1113; email: guralnic@math.usc.edu.

Titles in This Series

913 **Ethan Akin, Joseph Auslander, and Eli Glasner,** The topological dynamics of Ellis actions, 2008

912 **Igor Chueshov and Irena Lasiecka,** Long-time behavior of second order evolution equations with nonlinear damping, 2008

911 **John Locker,** Eigenvalues and completeness for regular and simply irregular two-point differential operators, 2008

910 **Joel Friedman,** A proof of Alon's second eigenvalue conjecture and related problems, 2008

909 **Cameron McA. Gordon and Ying-Qing Wu,** Toroidal Dehn fillings on hyperbolic 3-manifolds, 2008

908 **J.-L. Waldspurger,** L'endoscopie tordue n'est pas si tordue, 2008

907 **Yuanhua Wang and Fei Xu,** Spinor genera in characteristic 2, 2008

906 **Raphaël S. Ponge,** Heisenberg calculus and spectral theory of hypoelliptic operators on Heisenberg manifolds, 2008

905 **Dominic Verity,** Complicial sets characterising the simplicial nerves of strict ω-categories, 2008

904 **William M. Goldman and Eugene Z. Xia,** Rank one Higgs bundles and representations of fundamental groups of Riemann surfaces, 2008

903 **Gail Letzter,** Invariant differential operators for quantum symmetric spaces, 2008

902 **Bertrand Toën and Gabriele Vezzosi,** Homotopical algebraic geometry II: Geometric stacks and applications, 2008

901 **Ron Donagi and Tony Pantev (with an appendix by Dmitry Arinkin),** Torus fibrations, gerbes, and duality, 2008

900 **Wolfgang Bertram,** Differential geometry, Lie groups and symmetric spaces over general base fields and rings, 2008

899 **Piotr Hajłasz, Tadeusz Iwaniec, Jan Malý, and Jani Onninen,** Weakly differentiable mappings between manifolds, 2008

898 **John Rognes,** Galois extensions of structured ring spectra/Stably dualizable groups, 2008

897 **Michael I. Ganzburg,** Limit theorems of polynomial approximation with exponential weights, 2008

896 **Michael Kapovich, Bernhard Leeb, and John J. Millson,** The generalized triangle inequalities in symmetric spaces and buildings with applications to algebra, 2008

895 **Steffen Roch,** Finite sections of band-dominated operators, 2008

894 **Martin Dindoš,** Hardy spaces and potential theory on C^1 domains in Riemannian manifolds, 2008

893 **Tadeusz Iwaniec and Gaven Martin,** The Beltrami Equation, 2008

892 **Jim Agler, John Harland, and Benjamin J. Raphael,** Classical function theory, operator dilation theory, and machine computation on multiply-connected domains, 2008

891 **John H. Hubbard and Peter Papadopol,** Newton's method applied to two quadratic equations in $\mathbb{C}^2$ viewed as a global dynamical system, 2008

890 **Steven Dale Cutkosky,** Toroidalization of dominant morphisms of 3-folds, 2007

889 **Michael Sever,** Distribution solutions of nonlinear systems of conservation laws, 2007

888 **Roger Chalkley,** Basic global relative invariants for nonlinear differential equations, 2007

887 **Charlotte Wahl,** Noncommutative Maslov index and eta-forms, 2007

886 **Robert M. Guralnick and John Shareshian,** Symmetric and alternating groups as monodromy groups of Riemann surfaces I: Generic covers and covers with many branch points, 2007

885 **Jae Choon Cha,** The structure of the rational concordance group of knots, 2007

TITLES IN THIS SERIES

884 **Dan Haran, Moshe Jarden, and Florian Pop,** Projective group structures as absolute Galois structures with block approximation, 2007

883 **Apostolos Beligiannis and Idun Reiten,** Homological and homotopical aspects of torsion theories, 2007

882 **Lars Inge Hedberg and Yuri Netrusov,** An axiomatic approach to function spaces, spec tral synthesis and Luzin approximation, 2007

881 **Tao Mei,** Operator valued Hardy spaces, 2007

880 **Bruce C. Berndt, Geumlan Choi, Youn-Seo Choi, Heekyoung Hahn, Boon Pin Yeap, Ae Ja Yee, Hamza Yesilyurt, and Jinhee Yi,** Ramanujan's forty identities for Rogers-Ramanujan functions, 2007

879 **O. García-Prada, P. B. Gothen, and V. Muñoz,** Betti numbers of the moduli space of rank 3 parabolic Higgs bundles, 2007

878 **Alessandra Celletti and Luigi Chierchia,** KAM stability and celestial mechanics, 2007

877 **María J. Carro, José A. Raposo, and Javier Soria,** Recent developments in the theory of Lorentz spaces and weighted inequalities, 2007

876 **Gabriel Debs and Jean Saint Raymond,** Borel liftings of Borel sets: Some decidable and undecidable statements, 2007

875 **C. Krattenthaler and T. Rivoal,** Hypergéométrie et fonction zêta de Riemann, 2007

874 **Sonia Natale,** Semisolvability of semisimple Hopf algebras of low dimension, 2007

873 **A. J. Duncan,** Exponential genus problems in one-relator products of groups, 2007

872 **Anthony V. Geramita, Tadahito Harima, Juan C. Migliore, and Yong Su Shin,** The Hilbert function of a level algebra, 2007

871 **Pascal Auscher,** On necessary and sufficient conditions for L^p-estimates of Riesz transforms associated to elliptic operators on $\mathbb{R}^n$ and related estimates, 2007

870 **Takuro Mochizuki,** Asymptotic behaviour of tame harmonic bundles and an application to pure twistor D-modules, Part 2, 2007

869 **Takuro Mochizuki,** Asymptotic behaviour of tame harmonic bundles and an application to pure twistor D-modules, Part 1, 2007

868 **Gelu Popescu,** Entropy and multivariable interpolation, 2006

867 **Vilmos Totik,** Metric properties of harmonic measures, 2006

866 **William Craig,** Semigroups underlying first-order logic, 2006

865 **Nathanial P. Brown,** Invariant means and finite representation theory of $C*$-algebras, 2006

864 **John M. Lee,** Fredholm operators and Einstein metrics on conformally compact manifolds, 2006

863 **M. Lübke and A. Teleman,** The Universal Kobayashi-Hitchin correspondence on Hermitian manifolds, 2006

862 **Alberto Canonaco,** The Beilinson complex and canonical rings of irregular surfaces, 2006

861 **Leon A. Takhtajan and Lee-Peng Teo,** Weil-Petersson metric on the universal Teichmüller space, 2006

860 **Thomas M. Fiore,** Pseudo limits, biadjoints and pseudo algebras: Categorical foundations of conformal field theory, 2006

859 **N. Arcozzi, R. Rochberg, and E. Sawyer,** Carleson measures and interpolating sequences for Besov spaces on complex balls, 2006

858 **Enrico Valdinoci, Berardino Sciunzi, and Vasile Ovidiu Savin,** Flat level set regularity of p-Laplace phase transitions, 2006

For a complete list of titles in this series, visit the
AMS Bookstore at **www.ams.org/bookstore/**.